To Reverend Wilfred P. Schoenberg, S.J. who has preserved the heritage of the American Indian through his development of the Museum of Native American Culture at Spokane, Washington. We share great admiration for Adrien Voisin, the man, and Adrien Voisin, the sculptor.

Bill Harmsen

VISION

...and understanding of the lost-wax process of an ancient form of three-dimensional art dating back before the time of Christ—never fully understood as to the method, but always enjoyed and appreciated, which is the purpose of this book...

VOISIN

Adrien Alexandre Voisin...
Artist and Sculptor Par Excellence
Great American Recorder

1890—Born, Islip, Long Island, New York

1929—Named "AH-GAW-CHE-NENA," Honorary Member of the Blackfeet tribe

1931—Awarded the Diplôme d'Honeur by the French government

1932—Recipient of the Gold Medal at the International Art Exhibition in Paris

1979—Departed for the Happy Hunting Ground to Parley with his Brother, the Red Man, May 7, 1979

ILLUSTRATING THE LOST WAX METHOD

SCULPTURE TO BRONZE

FEATURING THE LIFE AND SCULPTURE OF

ADRIEN ALEXANDRE VOISIN

AUTHOR: BILL HARMSEN

PUBLISHED BY
HARMSEN PUBLISHING COMPANY
3131 EAST ALAMEDA AVENUE, SUITE 2001
DENVER, COLORADO 80209

FIRST PRINTING 1981

ALL BRONZES SHOWN AND LISTED IN THIS BOOK ARE FROM THE HARMSEN WESTERN AMERICANA COLLECTION

LIBRARY OF CONGRESS CATALOG CARD NUMBER 80-84752

CONTENTS

INDEX

INTRODUCTION

America now has more serious art collectors than ever before in its history. Within the private collections in this country there is to be found a wealth of museum-quality art, antiques and artifacts of unlimited diversity. Our museums, in most cases, have been founded around a nucleous of an outstanding collection, given by a devotee of the arts. There is much to be said for the collectors, for they are the people who make the museums possible.

Responsible collecting is the result of two things — good advice from knowledgeable people, and many hours of study, which invariably results in a very rewarding adventure.

My first serious collecting involved oil paintings and water colors. I wanted one good example of an artist's work, and the paintings I looked for generally were from an artist about whom I had studied. Knowing the history and background of the artist adds another dimension — call it the human touch. I would seek out members of the National Academy of Design; ethnologists; topographical specialists who accompanied various expeditions into the New Frontier; members of the Hudson River School and the "Ashcan" school; the Taos and Santa Fe groups, the Rocky Mountain School; the WPA artists; and many of the great contemporary artists that I visited with personally during meetings at various art symposiums and shows.

Sculpture did not have an appeal for me at first. I had the preconceived idea that if there were twenty castings of a bronze, it was a production run.

I remember the time I visited Olaf Wieghorst at his home in California, hoping to buy one of his paintings, but his wife, May, said they were going to keep all of the paintings that they had displayed, as they were her favorites.

During my visit there, Olaf, knowing that I had raised and ridden quarter horses, took me out to the barn to see his favorite mount — a great quarter horse; good straight legs, nice gaskin muscles, fine head, and small ears. I was impressed. When we went back to the house to have coffee, I observed a small bronze of a cowboy and horse modeled after his own fine quarter horse, on the table. Olaf said, "Bill, that's number 20 of 20. I'm keeping number one. The rest are gone. I know you like horses. I would be glad to sell you this one . . ."

Then he proceeded to give me an informative, two-hour lesson on the work that goes into each casting and proved to me that they are not "stamped out." Well, my education on casting a piece of sculpture was started. My outlook on this type of art was changed, and my fervor for collecting bronzes was stimulated.

In the next two years I acquired bronzes by over twenty different sculptors — all great! — some deceased, and many, contemporary.

It was in February of 1977, when my wife, Dorothy, and I attended the Art Auction at Spokane, Washington, sponsored by the Museum of Native American Culture. It was there we met Father Schoenberg, the art-loving Jesuit priest from Gonzaga University, who had the vision to put together this great museum.

He had acquired part of Adrien Voisin's life work, and has it on permanent display. One of the large bronzes caught my eye — "The Happy Hunting Ground." After I learned the history of the piece, I knew it would be an important addition to our collection. Father Schoenberg's enthusiasm for this artist easily brushed off on Dorothy and me. He subsequently suggested that on my next trip to the Los Angeles area, I should visit the fragile 84-year-old Adrien Voisin.

This I did, through the aid of the artist's sister, Marguerite Baker, as Voisin's wife was quite ill at the time. I will never forget my first visit with this humble, aristocratic man. It was a real pleasure. At the outset, I found it was difficult for him to talk about himself. His conversation was always about others — especially the Indians he had known and loved and sculptured. He showed me the beaded pipe bag with the peace pipe that was given to him by Two Guns White Calf when he, Voisin, was made an honorary member of his tribe, and also a pair of moccasins from "Mountain Chief," among other things.

As we talked, I realized how great Voisin really was — his philosophy on life, his love for animals, and his compassion for all men impressed me. On subsequent visits, I did get him to talk about himself, and he would smile a little or laugh out loud when I played back the tapes I had recorded of him telling me about his life.

Eventually, I arranged to acquire many pieces of sculpture from Voisin that he had not cast. Some were so brittle from age that the plaster disintegrated when separated from the molds that I had made. His favorite works were three large, bas-relief slates. He explained the many trial and error pieces that he had worked on and how he finally accomplished what had never been done before by any of his predecessors. These pieces were never out of his home and studio in San Francisco until he moved with his wife, Mary, to Los Angeles. He built sturdy crates, lined them with protective carpeting, crated the pieces, and had his nephew take them to his sister in Los Angeles for safekeeping.

I remember my first view of these. Here was sculpture on slate, a completely different medium from anything I had ever seen. And a big day in my collecting came when Adrien wrote to say he had decided to sell these pieces to me because he was certain I would take proper care of them and have them cast in bronze to preserve them for future generations.

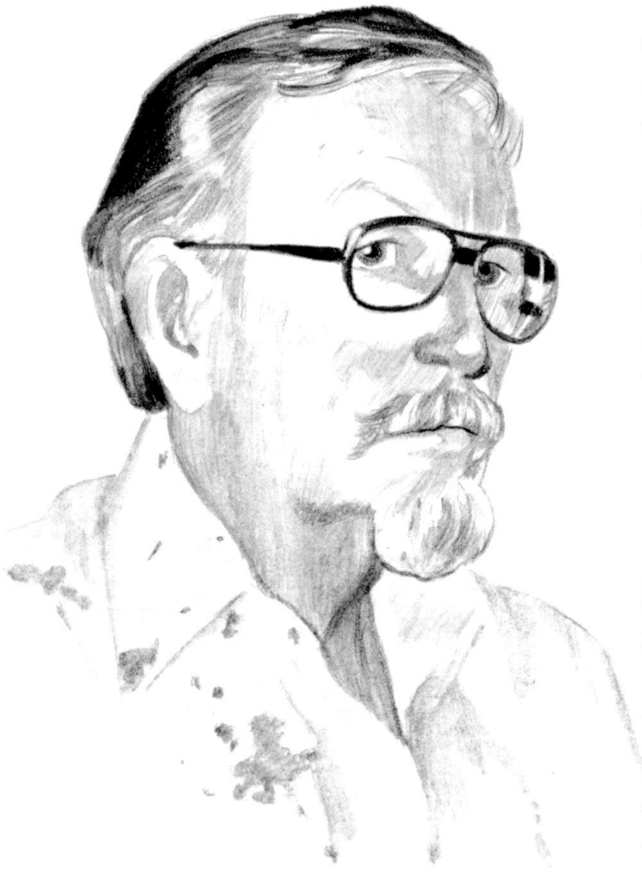

BILL HARMSEN

The one thing I learned from Voisin was the realization of how complicated, tedious, and painstaking a process it is for a sculptor to accomplish a finished piece of work.

In talking with collectors at art shows around the country, I came to one conclusion; most people are unaware of the lost-wax process of casting and like myself at one time, think sculpture is easily produced.

Because bronze sculpture is almost immortal in nature. It is not my right to assume that it is an easy expression of the artist's talent, it never has been and never will be. Many sculptors have had to find renewed inspiration in their work by turning from a no-color substance such as plaster, wax, and bronze or stone, to easel painting for a short period, thus relieving the pressure of working with heavy masses for long periods and becoming buried in a drab world of grays and browns.

I hope this book will clear up some of the misconceptions, and will evoke more appreciation of this unique and important art form.

THE PRIMITIVE BEGINNING OF AMERICAN

SCULPTURE

One never really stops to think of how a custom or a craft started in America. Some have assumed that sculpture was a nineteenth-century art form; others believe it was introduced by an adventurous English artist arriving on the American scene to capture the beauties of a new land; or some may have thought sculpture too unimportant to even wonder about. At any rate, it became an ignored form of artistry. But it actually has been around for a long time and will be for some time to come.

Classic and contemporary sculpture is quickly becoming one of the best investments in America today, as the prices of the Fine Arts have more than tripled in the past decade.

The humble beginning of sculpture in America was in the seventeenth century on gravestones, wooden Bible boxes, and utilitarian objects. The first 125 years of American sculpture are found mainly on the slate gravestones that are in the church graveyards from Maine to South Carolina.

Contrary to popular conception, the first colonists were of the middle or lower classes—merchants, ministers, craftsmen, laborers, and vagabonds. Artisans were not in evidence among this group.

The need for burial markers arose in the colonies. Hence, until the time of the American Revolution, sculpture in stone was accomplished by a local stonecutter who used slate or limestone and engraved it with names and dates. Later in the century, bas-relief designs were introduced, and the gravemarkers became more elaborate. By 1740, native-born artisans began to create portrait busts and statues in three-dimensional, sculpturesque form.

In the colonial period, another form of sculpture was developing. That was the product of the woodcarvers, whose talents were limited to the designs incised on the surfaces of the furniture they made. In the 1730's, cabinetmakers and artisan carvers began to copy design books from England. It was in the shops of these gifted, unschooled craftsmen that America's first true sculptural art had its origin and the first "native school" of sculpture had its beginnings. The shipyards also produced master woodcarvers and carvings that adorned the sailing ships, which proudly displayed massive figureheads and decorative stern boards.

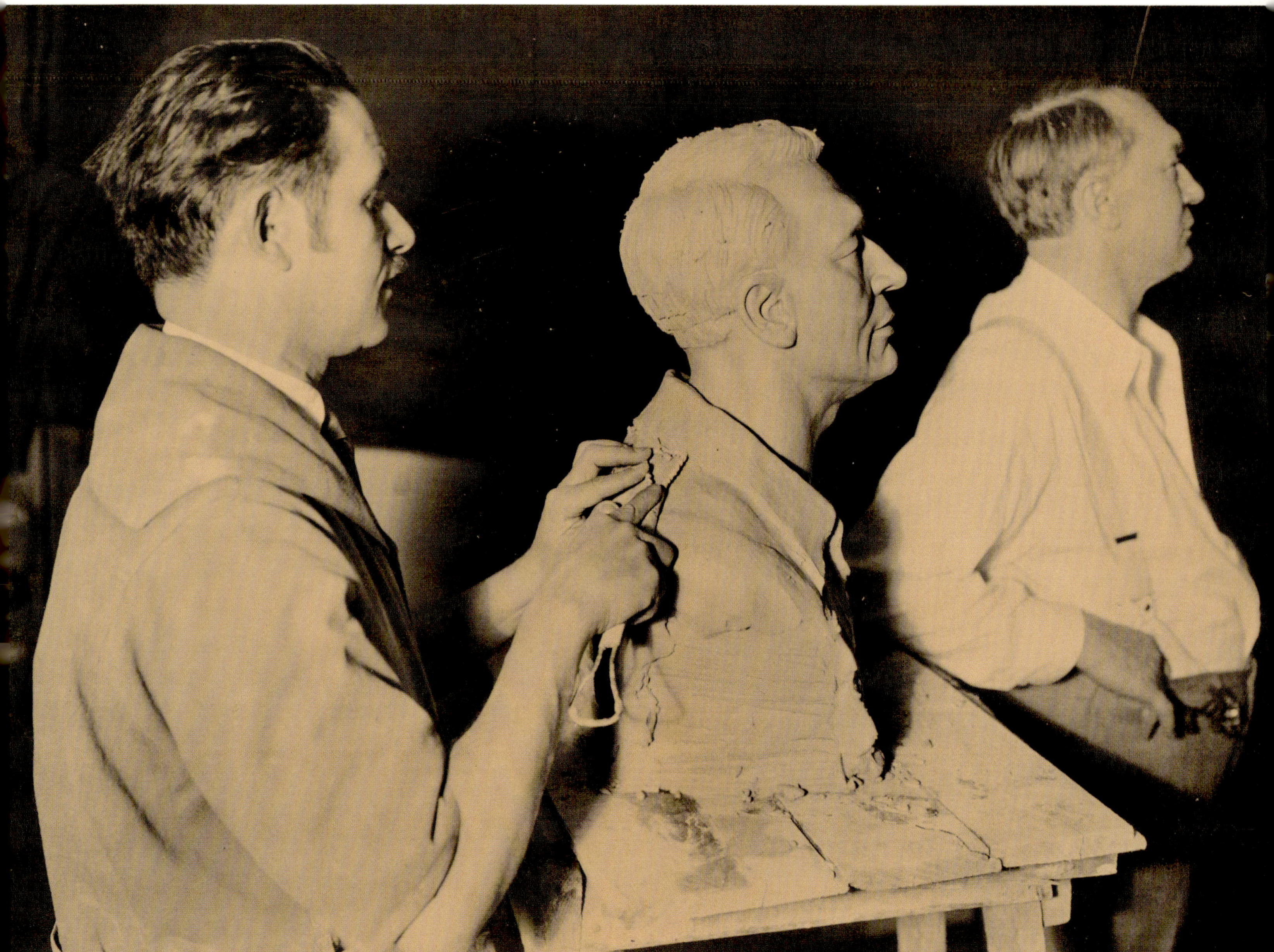

Adrien Voisin sculpting Vachel Lindsay, famous poet and author of "Abraham Lincoln Walks at Midnight," and "The Ghosts of the Buffaloes."

These craftsmen contributed greatly to American folk art in the forms of tavern signs, standing figures of cigar-store Indians and sailors, circus-wagon figures, weather vanes, duck decoys, children's toys, and treen ware. (Treen ware is a coined expression for kitchen utensils made from the wood of a tree.) Though unschooled, these gifted woodcarvers pursued their crafts for the commercial trade right up to the 1840's.

The middle of the eighteenth century saw some changes from folk-art motifs to the use of more classical designs in architecture and furniture—a direct European influence—and continued until after 1835, when woodcarving per se began to give way to the more sophisticated forms of sculptural endeavors.

William Rush, 1756-1833, did much to make Philadelphia known for some of the finest early, carved furniture. Also, he was one of the first to attempt sculpturing in clay or terra cotta and plaster.

After his death, modeling in wax, which is the initial step in bronze casting, was introduced into America. Wax, being in great supply in the colonies, soon became a medium used by men and women artisans alike, from bas-relief portraits, small figures and tableaux to the more ambitious figures in wax museums. Surprisingly, much of the early wax work has survived. Some of the better wax art forms were sent to Europe for casting in bronze.

The first-known sculpture in bronze was cast in America in the 1850's by Robert Ball Hughes. It was placed in Mount Auburn Cemetery near Cambridge, Massachusetts. The statue was cast in Boston in the sculptor's own foundry at a cost of $3,293.78, but it deteriorated so rapidly that it had to be recast in a foundry in Paris. Bronze casting in America has traveled a long and arduous road in the last 125 years. Perhaps that is the reason it has been overlooked in a way it never will be again.

Excellent and mediocre sculptors are becoming very prolific in this age of aesthetic awareness. Our American foundries that are casting bronzes are finding techniques far superior to those of our European competitors. This fact helps to make the procedure a Labor of Love, when a sculptor can oversee each intricate step and follow the development of a piece that becomes so much a part of the artist in the delicate process of developing his Thing of Beauty.

The knowledge and interpolation of the lost-wax process of bronze casting will surely cultivate an increased and more profound interest and appreciation of this misunderstood form of art.

FROM WAX TO BRONZE BY THE LOST WAX METHOD

LOST WAX

Sculpture made from the "lost-wax," or vanishing-wax, method is perhaps the least understood of all the art forms produced today. To fully comprehend this ancient art, a person would have to put in much time and effort with an artist at a foundry observing the intricate procedures that must be undertaken.

This article is not intended to make a professional or expert out of the reader. Instead, an attempt is made here to present a comprehensive, elementary look at what takes place from the time the original model is formed by the artist to the point where the sculpture emerges complete.

To make the study easier to assimilate, an artist has made schematic drawings that will outline the procedures as we go.

Let's first look back into the history of this ancient art. It has been known to man for over 6,000 years, and because of its durability, has survived all other art forms. One can find early masterpieces formed in bronze, silver, and gold made by the lost-wax method in museums around the world, but the earliest artisans using the method appeared in China, India, Afghanistan, Egypt, the Near East, and in Italy and Greece. Today, the same technique is used, and with scientific development, the quality is much improved.

In the early days, ornate designs were made from the sand-casting method, which is still popular today and which was used a great deal by the Navajo Indians in making much of their jewelry. At present, however, 90% of all bronzes are made by the lost-wax method. This term is derived from the process itself, as the wax is actually lost.

1 Plaster Elk

For illustrative purposes, we will first examine the original model of THE FAWN ELK, which is small and "plain" and will be easier to follow and to understand than a large and more complicated form. Voisin sculptured this piece in clay in 1936.

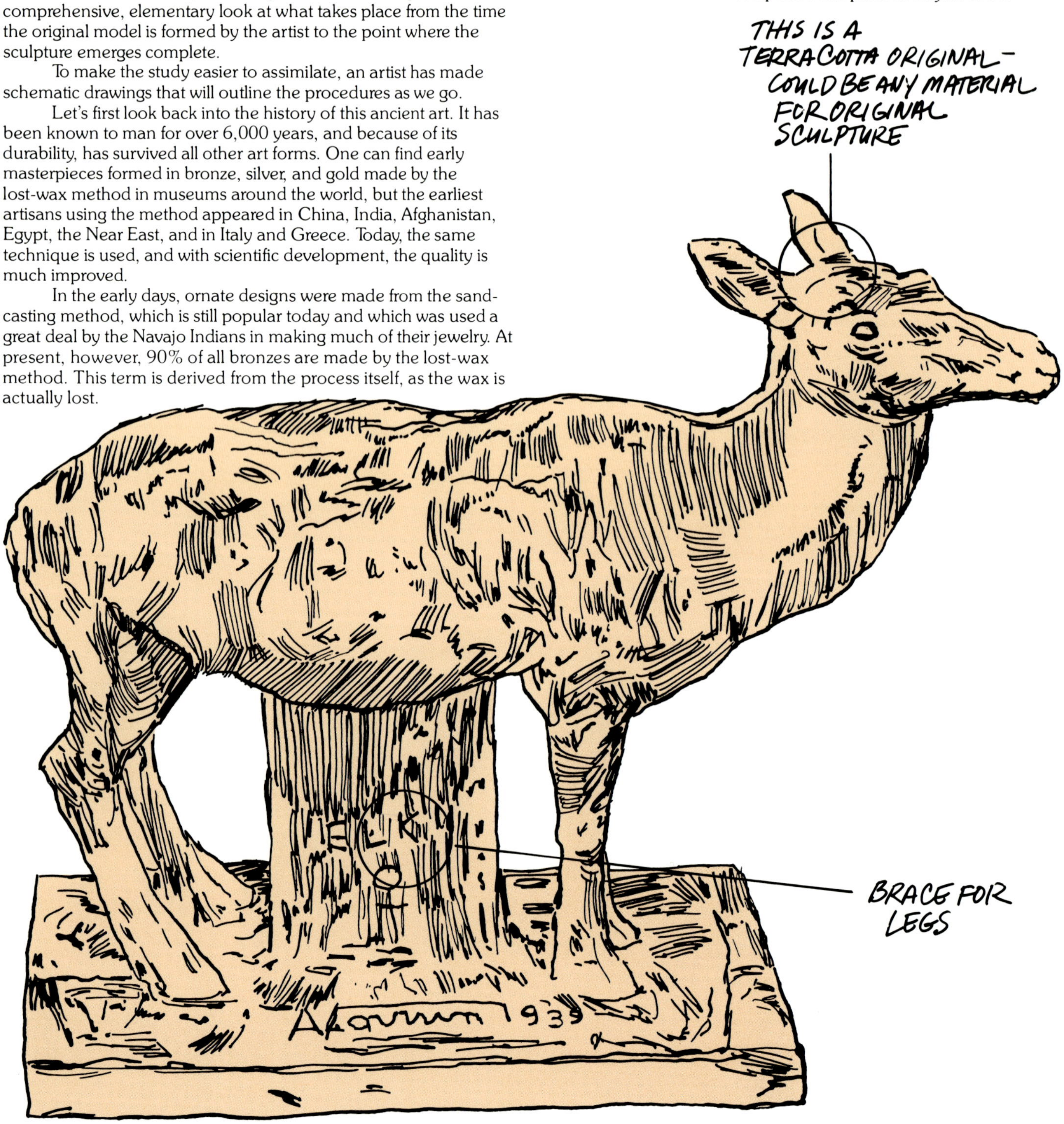

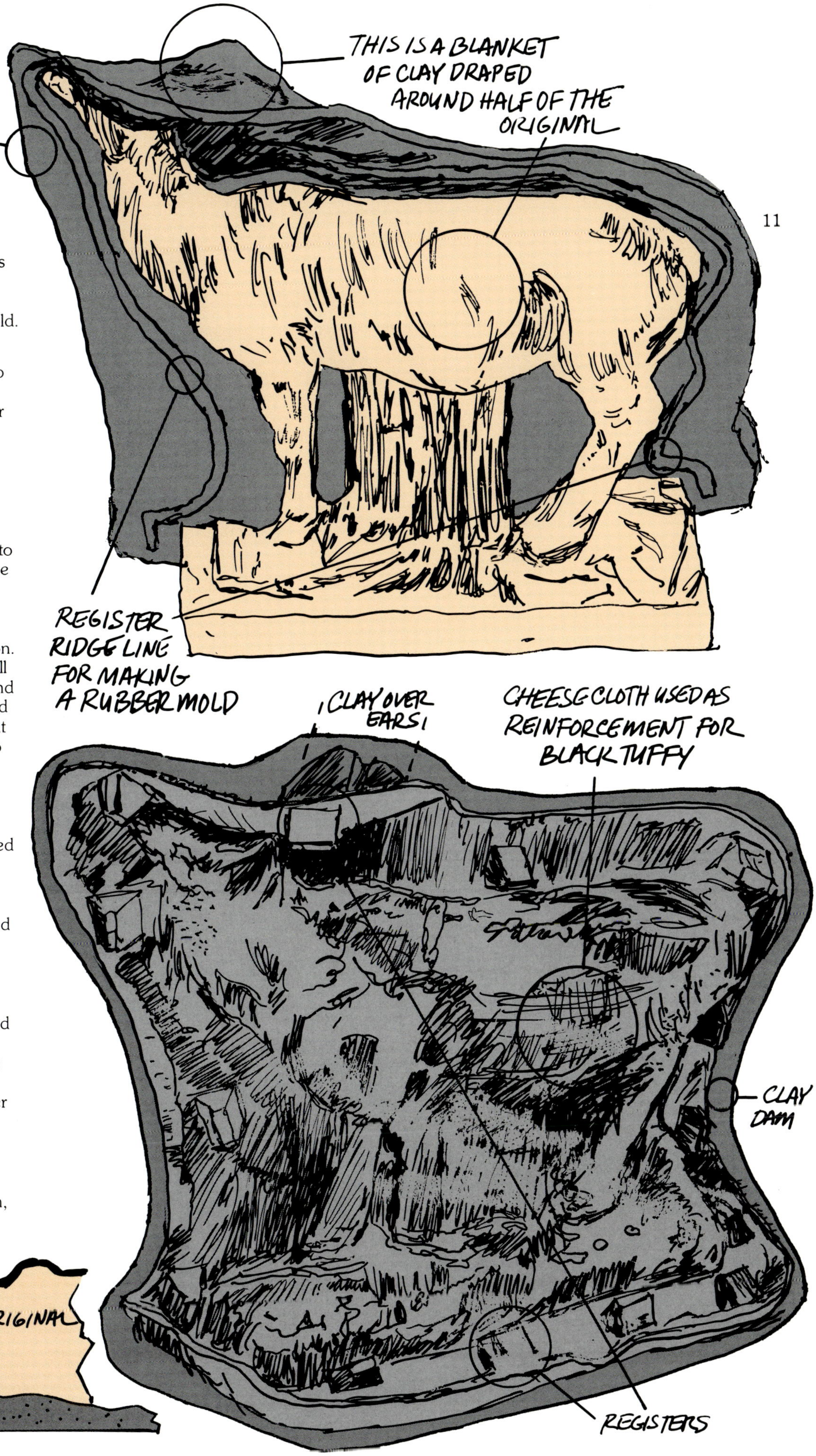

2 Rubber Mold

From this original, one of the first steps of the "Sculpture to Bronze" process is the molding of the original art work. There are many molding techniques; two of them are the **blanket** and the **pour** mold.

BLANKET MOLD: For a blanket mold, the original art is sectioned into two halves by chosing a common center line over the entire piece; and once this center line is established, of modeling clay or plasticine a clay wall is laid up to the first section to be molded, and the "cold molding compound" (CMC) is then painted on this half. The first layer is the "detail coat," meaning that it will pick up all the surface detail. This coat is allowed to "cure," and then several more layers of the CMC are applied for added strength.

When the final application of the CMC has cured, a "mother mold" of plaster of paris is applied to the first section. The piece is then turned over, the clay wall that is next to the sculpture is removed, and the same steps are repeated on the second section. In this mold are formed alignment keys or notches to properly adjust the two halves together.

POUR MOLD: This is a more modern method which is being used by many moldmakers today.

Once again, the art work is sectioned into two halves by the use of a common center line. Sometimes the work will require more than two sections for molding, particularly if it is a more involved sculpture. This time, the moldmaker "paints" a thin line on the sculpture as to where the dividing line will be for easy location later. A thin sheet of saran is wrapped around the sculpture and secured with cellophane tape. This is for easy removal of the two mold halves. Clay that has been pounded or rolled out to a quarter-inch thickness is then molded over the art work. A larger, thicker section of clay is build up over the predetermined dividing lines that were painted on the art work. This will serve as a division for the plaster halves and a thicker cutting section, for dividing the rubber mold.

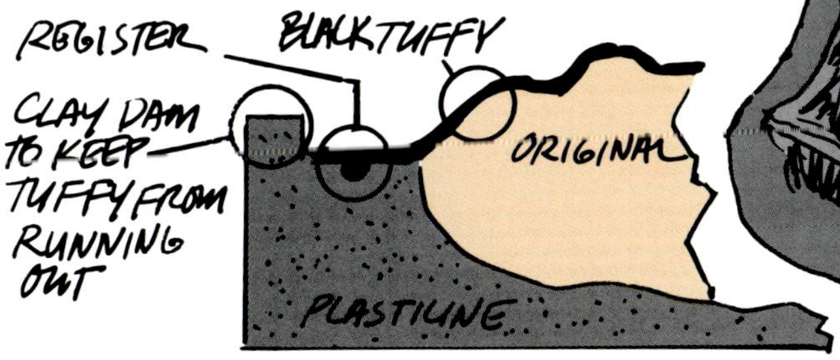

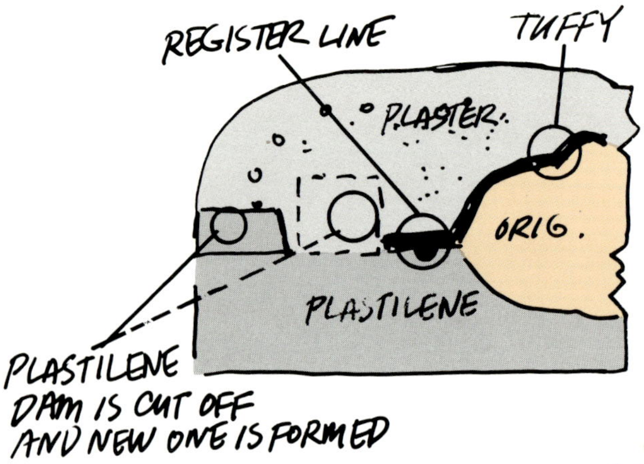
REGISTER LINE
TUFFY
PLASTER
ORIG.
PLASTILENE
PLASTILENE DAM IS CUT OFF AND NEW ONE IS FORMED

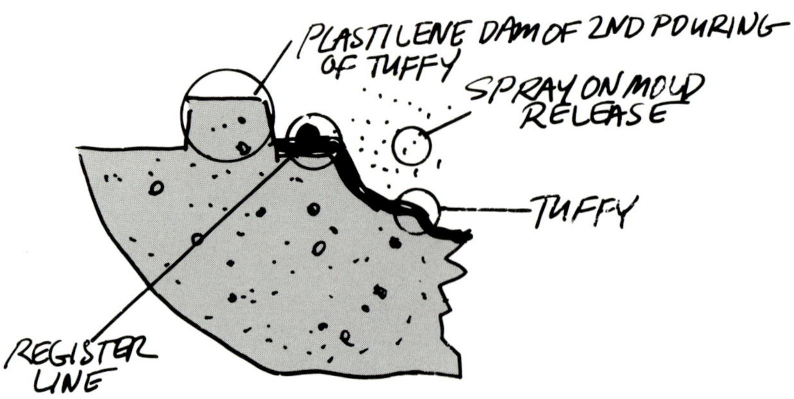
PLASTILENE DAM OF 2ND POURING OF TUFFY
SPRAY ON MOLD RELEASE
TUFFY
REGISTER LINE

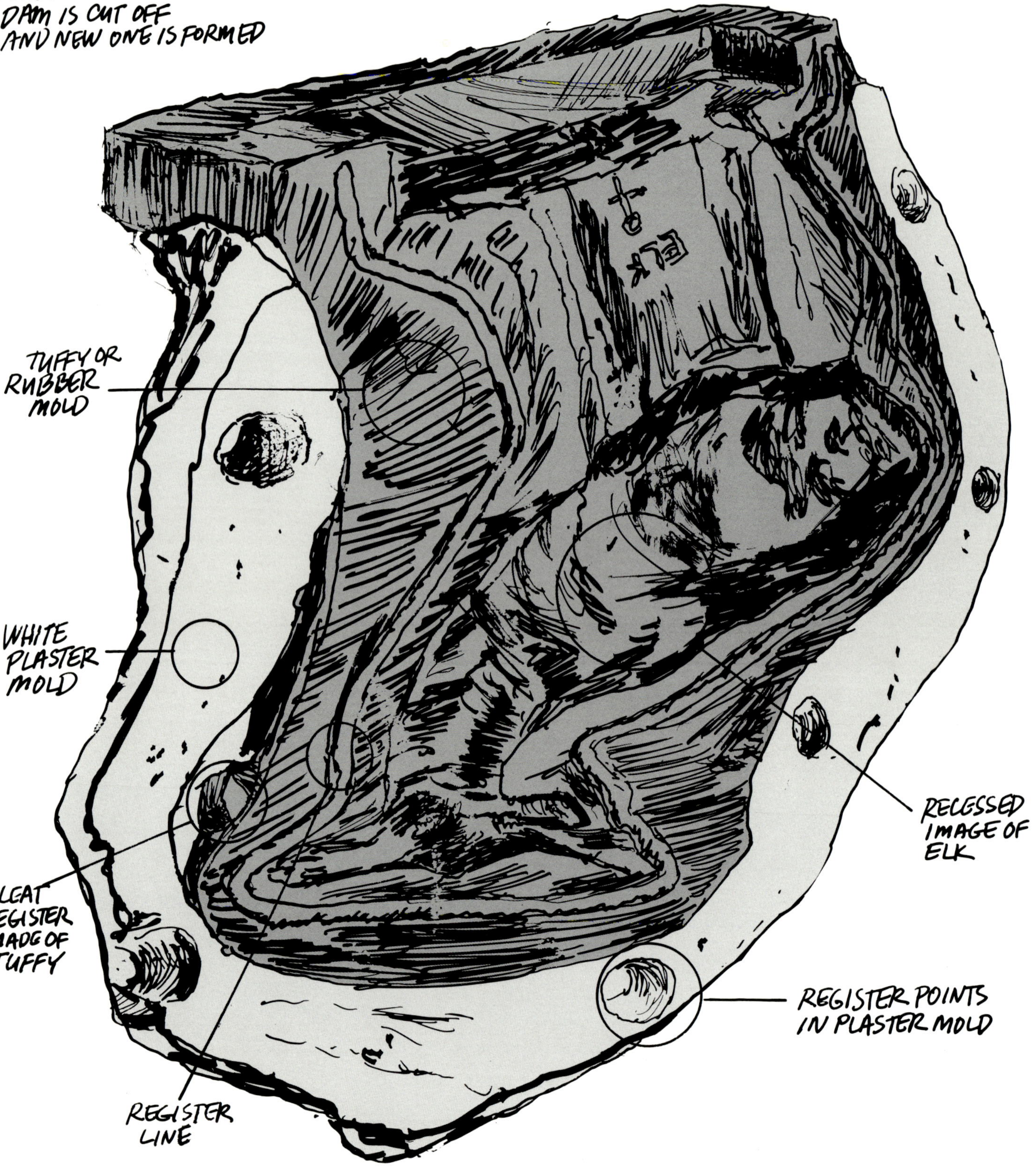
ELK
TUFFY OR RUBBER MOLD
WHITE PLASTER MOLD
CLEAT REGISTER MADE OF TUFFY
RECESSED IMAGE OF ELK
REGISTER POINTS IN PLASTER MOLD
REGISTER LINE

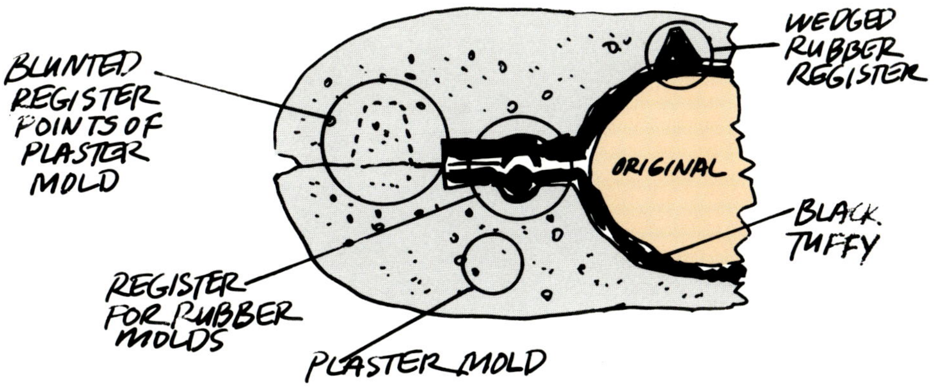

3 Plaster and Rubber Molds

Plaster of paris is then applied to the clayed-up sections of the mold. After the plaster has dried, it is removed from the clay. The clay, in turn, is removed from around the art work. At this point, the saran wrap is removed. The plaster sections are then placed back around the original art work. There will be a quarter-inch negative space around the original where the clay once was. The moldmaker then will pour CMC through a pour spout in the mold. The rubber will flow into the space that was left by the clay.

The rubber is allowed to cure, and the plaster mother mold is removed. The original art work is encased in the rubber now so that the moldmaker can open the rubber mold along the built-up edges. The rubber is cut open with a scalpel or sharp instrument in a zigzag or serrated pattern, which gives accurate alignment to the two halves; and the white line that was painted on the original art work in the beginning makes the cutting operation more certain.

After the rubber mold has been cut, the sections are placed in the plaster mother molds. These molds will be used each time a wax is made.

4 Pouring the Wax

Our next step is the forming of the wax, which will also be an exact duplicate of the original model. The two halves of the molds (rubber inside and plaster outside) are placed and locked together; then in an inverted position, the heated wax in liquid form is poured into the mold cavity. As the material hardens from the mold surface toward the center, the excess is poured out. Timing and temperature again are of the utmost importance, for it is necessary to pick the right moment to pour out the excess wax to form an exact duplicate with a wall thickness of one-eighth inch.

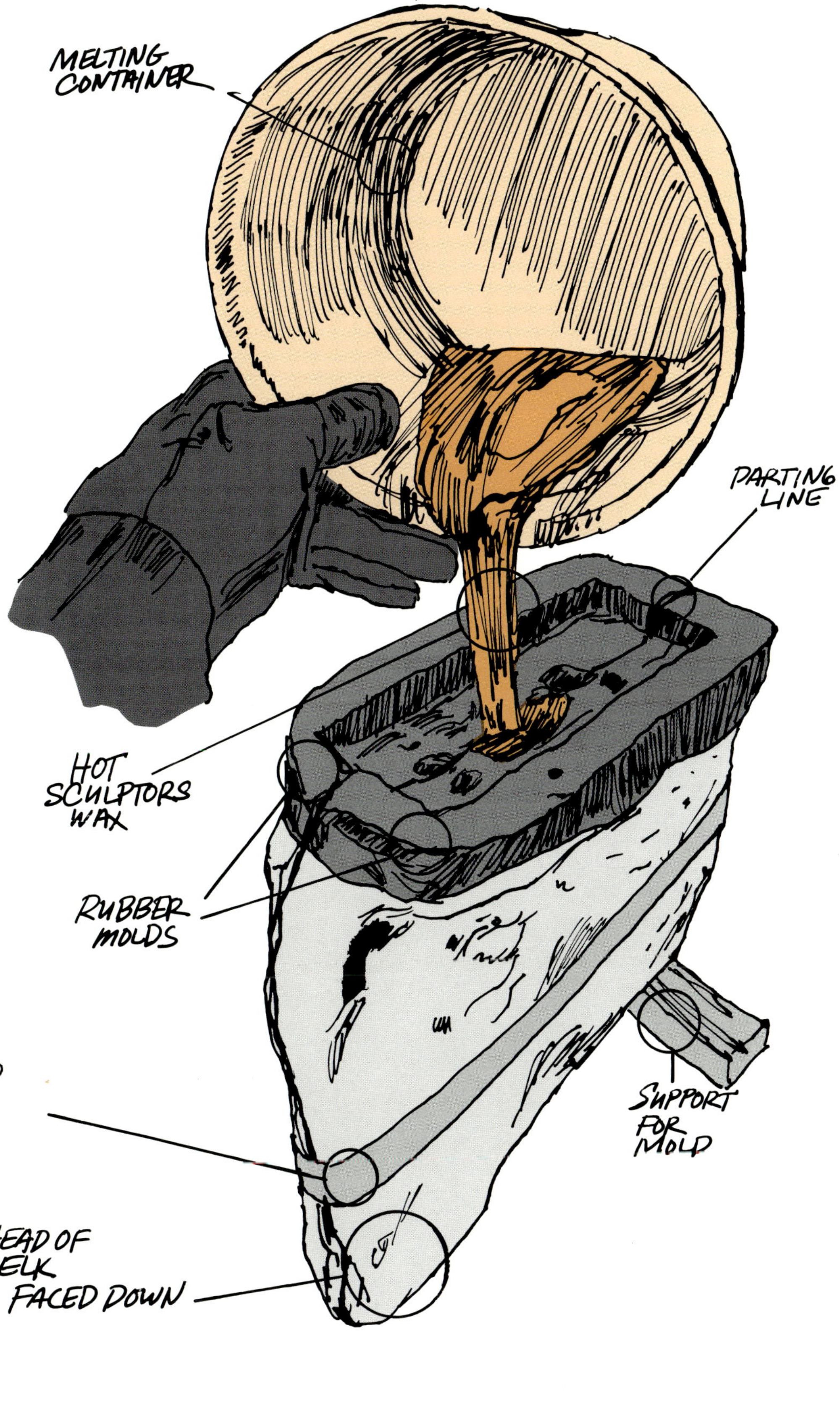

5 Removing the Mold

After the wax is set up and cooled, the plaster mother mold is removed, and the rubber mold is carefully peeled from the hollow wax replica.

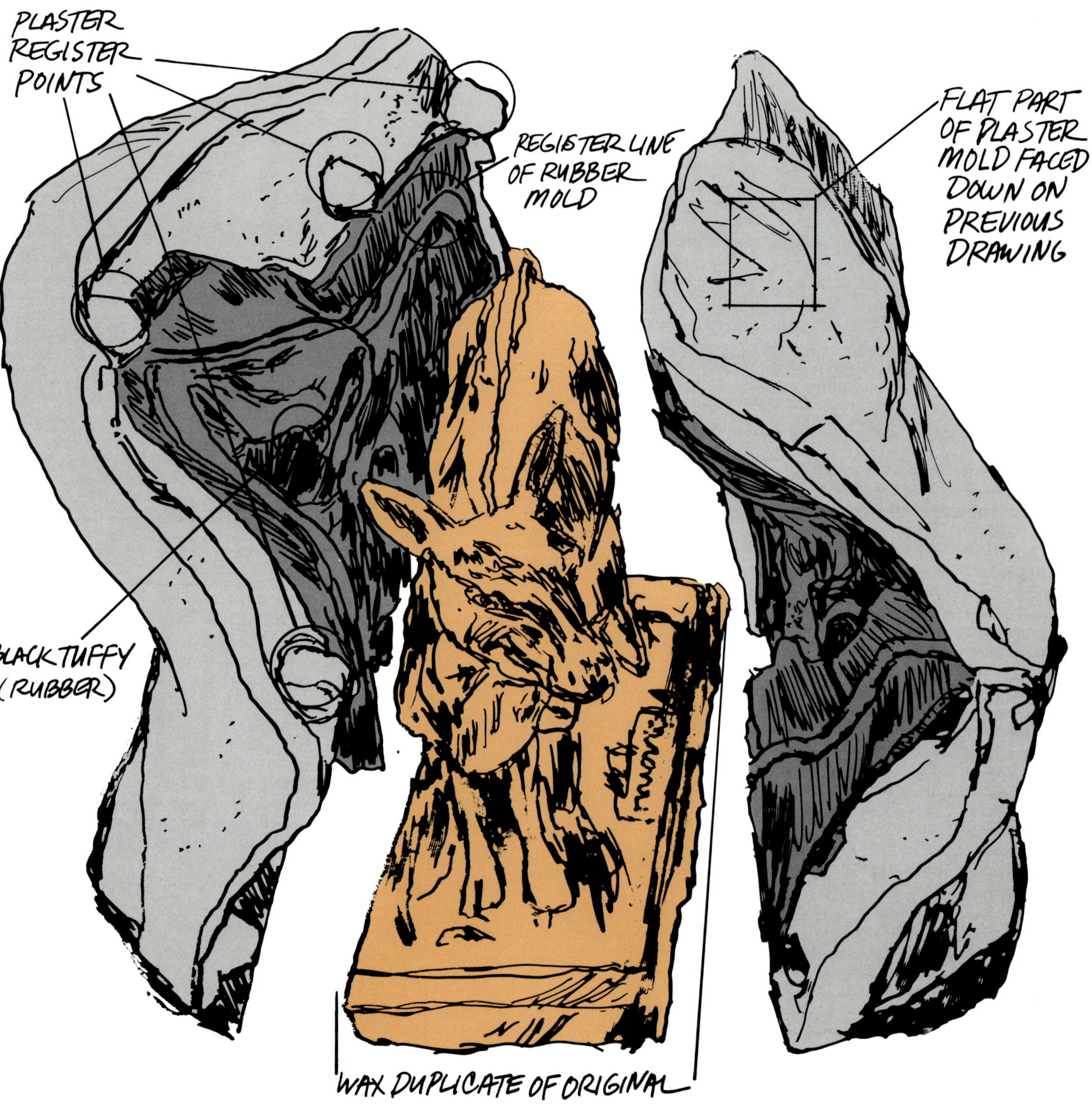

6 Chasing the Wax

The wax reproduction is carefully chased and prepared for attaching the gating and sprews necessary for the metal delivery system and venting.

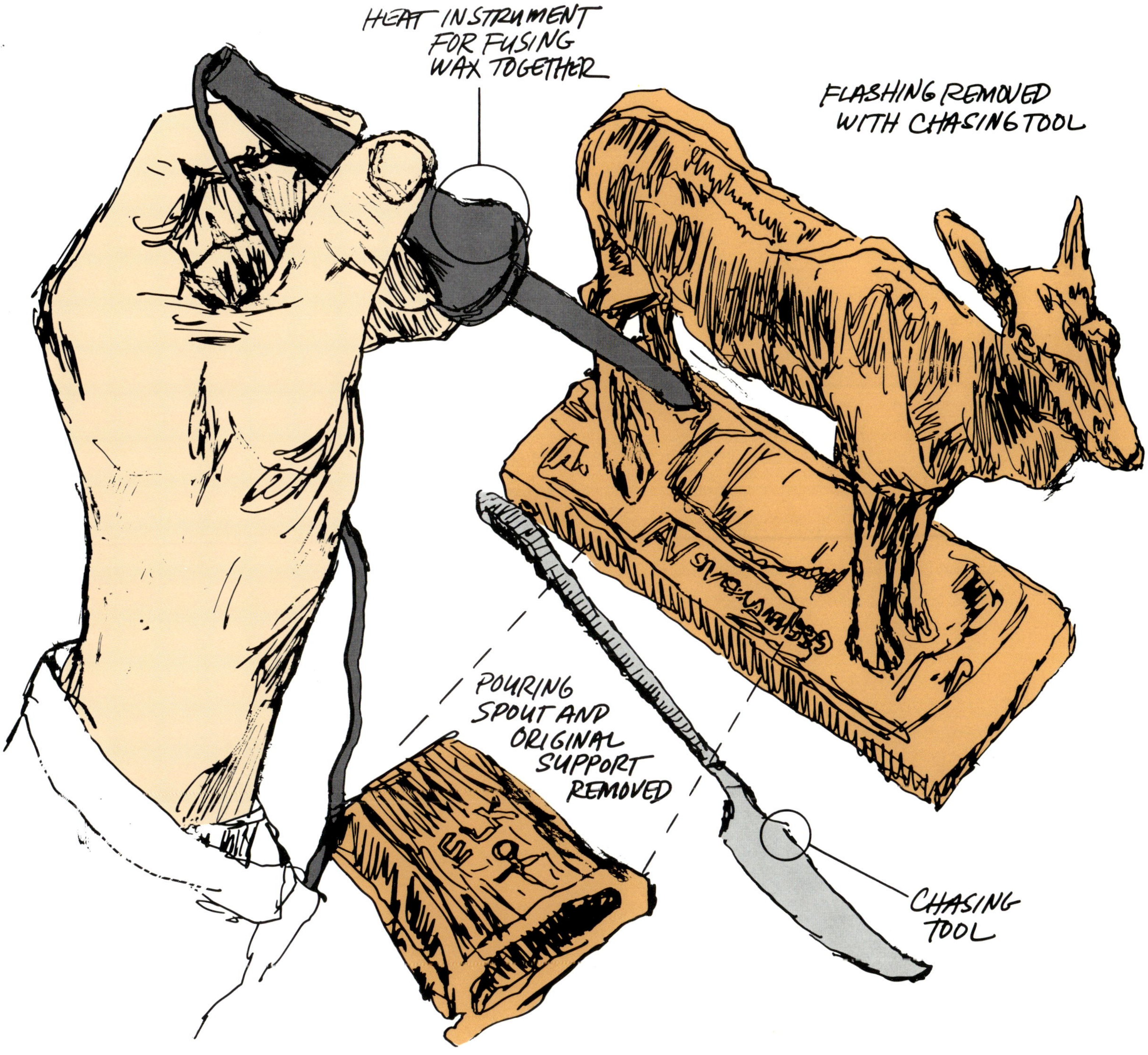

7 Ceramic-Coating Procedure

The wax pattern, properly prepared and attached to the casting sprue, which is also wax, is now ready for the ceramic shell to be started. It is dipped into a silicone slurry from five to eight times, depending on the size and type. Between each dipping, sand is sprinkled on the form. The first two dippings are sprinkled with 120 grain sand which is for detail in the sculpture. The following dippings are of 30 grain sand to give it strength and make it permeable after curing. It is important between each one of these steps to have proper curing and drying; otherwise it could end up with a weak shell, causing a failure in the metal casting process.

The drawing shows a section in the body of the elk cut out. This permits the entry of the slurry into the hollow of the wax, for it is important to ensure the production of a thin-wall casting. (Metal shrinks as it cools, and the thinness of the wall is important).

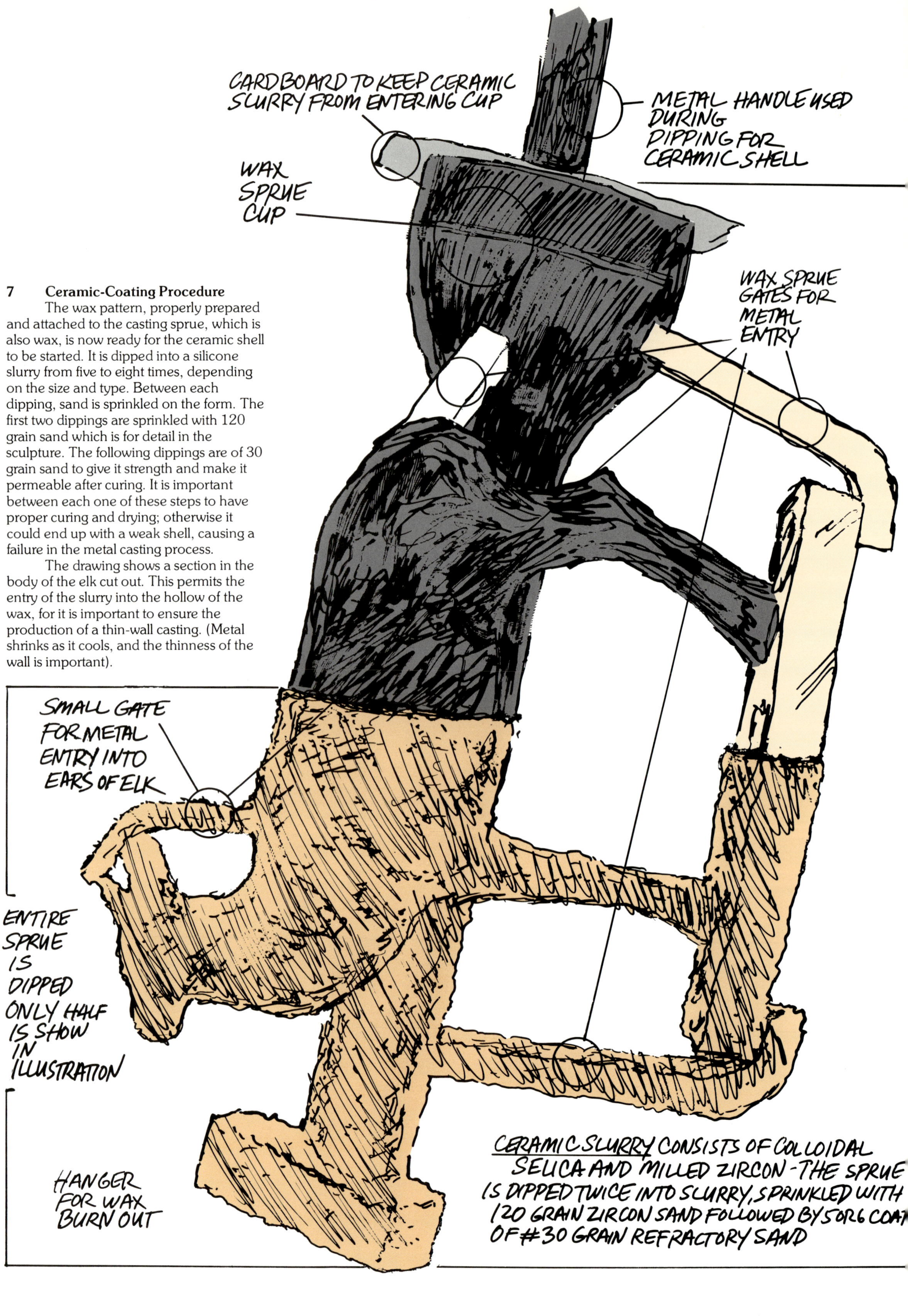

8 Melt-out—Lost Wax

At this point, the ceramic-filled mold is placed into a burn-out oven. This is the stage of the "lost-wax" procedure when the wax is actually melted out, leaving the hollow mold clean, dried, and cured. Some foundries add another step at this point by first heating the shell in an autoclave where a generator introduces steam under pressure at approximately 300°. The steam melts out approximately 98% of the wax, and then the shell is again heated in a dry-heat chamber to complete the melt-out, cure, and harden the investment.

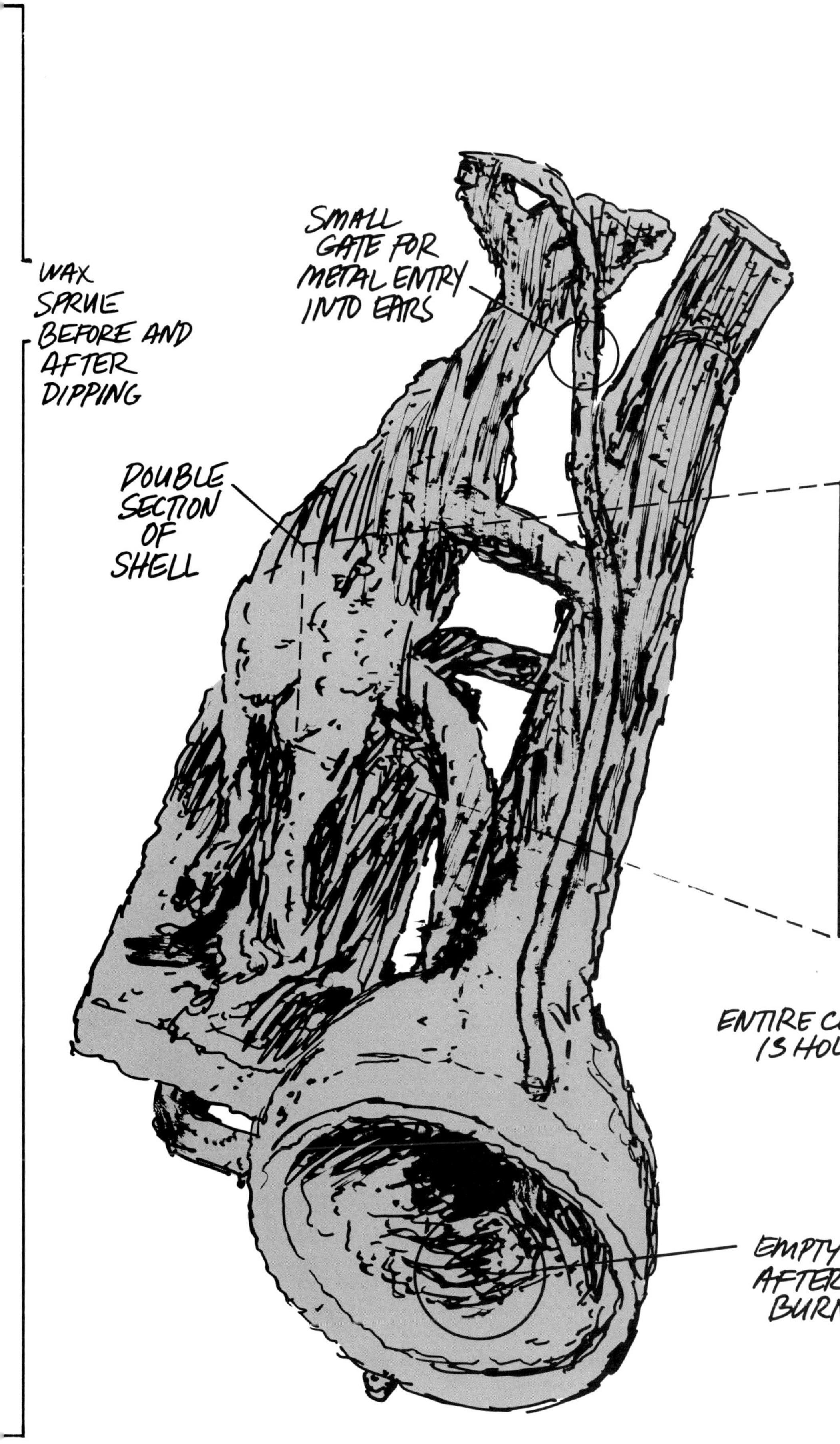

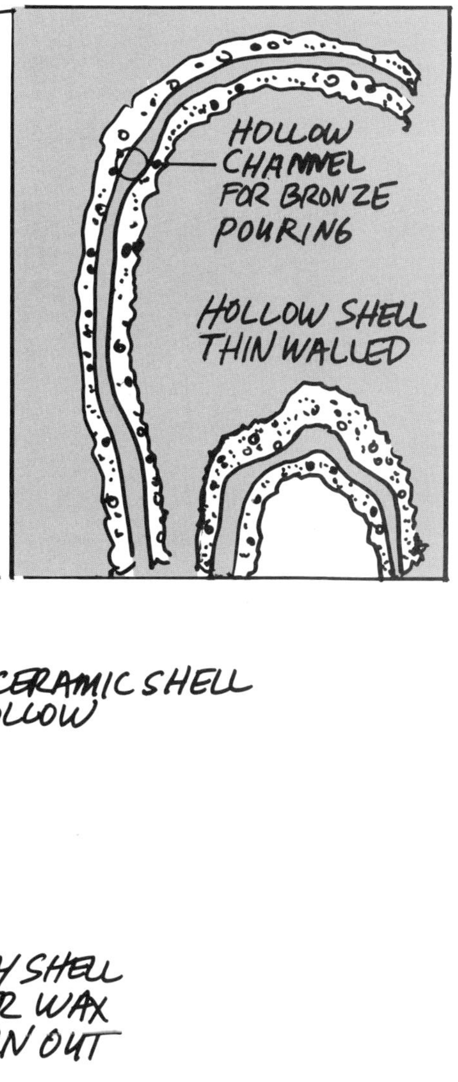

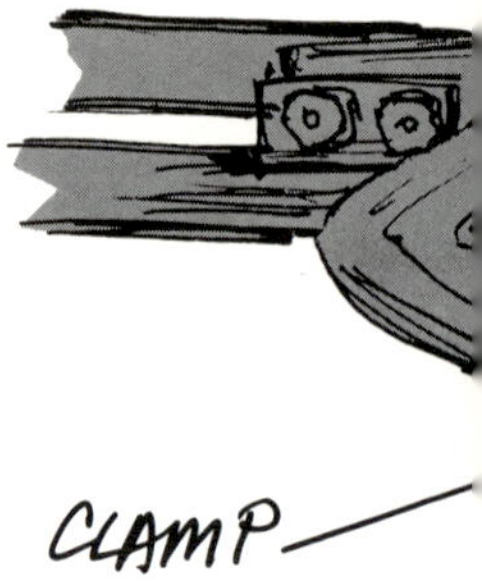

9 Positioning the Ceramic Mold

With the mold readied, it is put into an oven and heated to approximately 1400° and then placed on a pouring table or buried in a container of sand with the pouring cup up. This is a safety feature, for if the mold should break, the hot fluid bronze would be contained. It also adds strength to the mold.

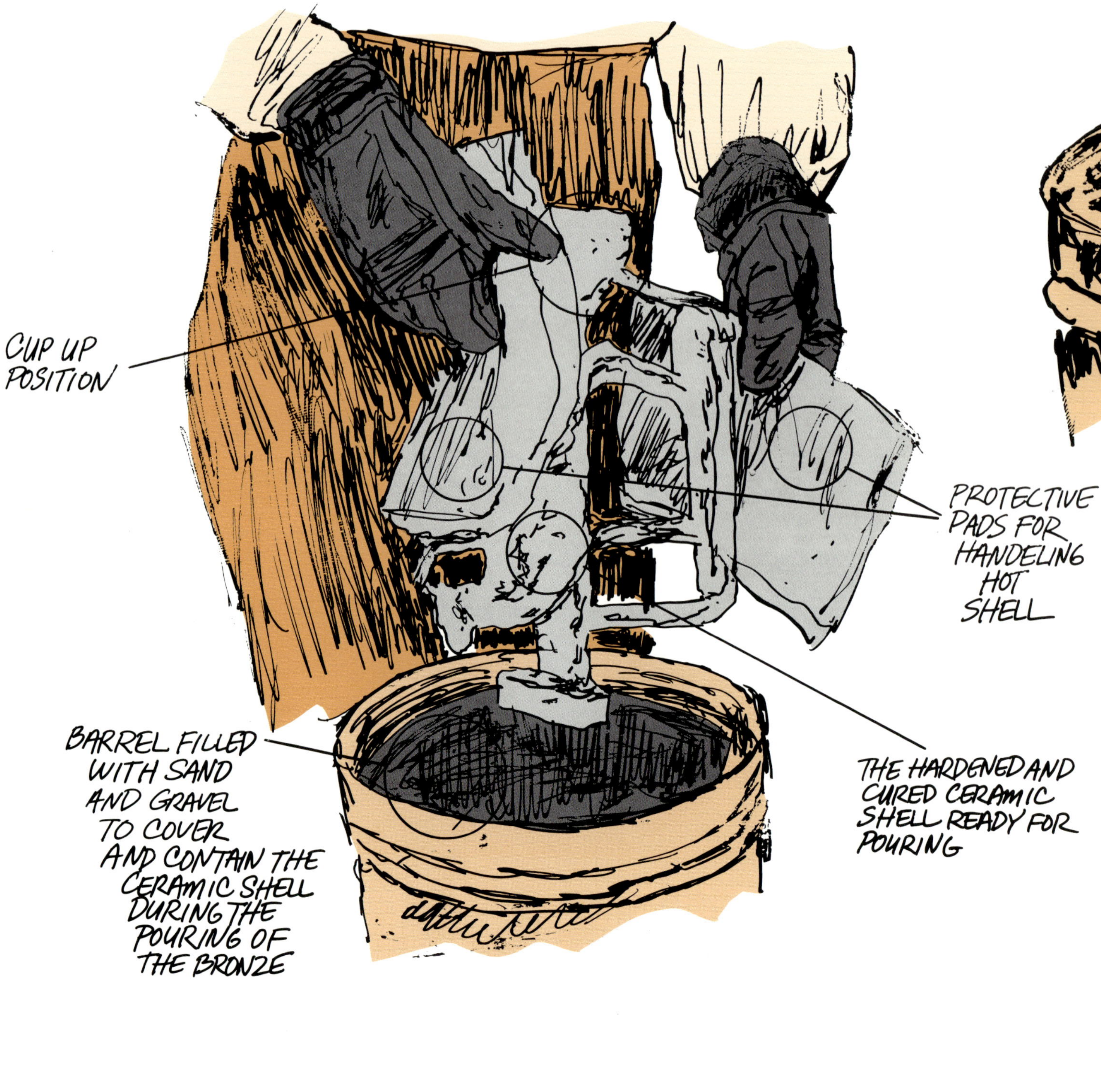

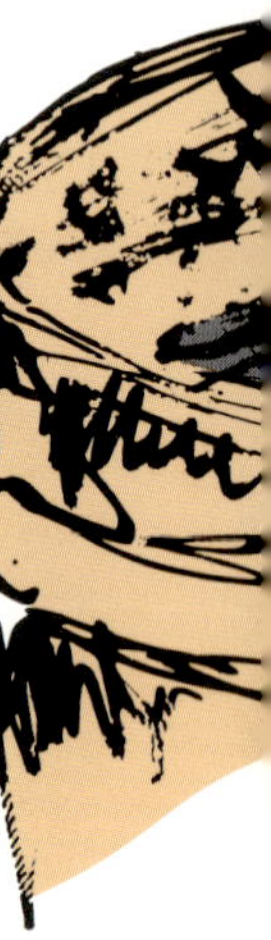

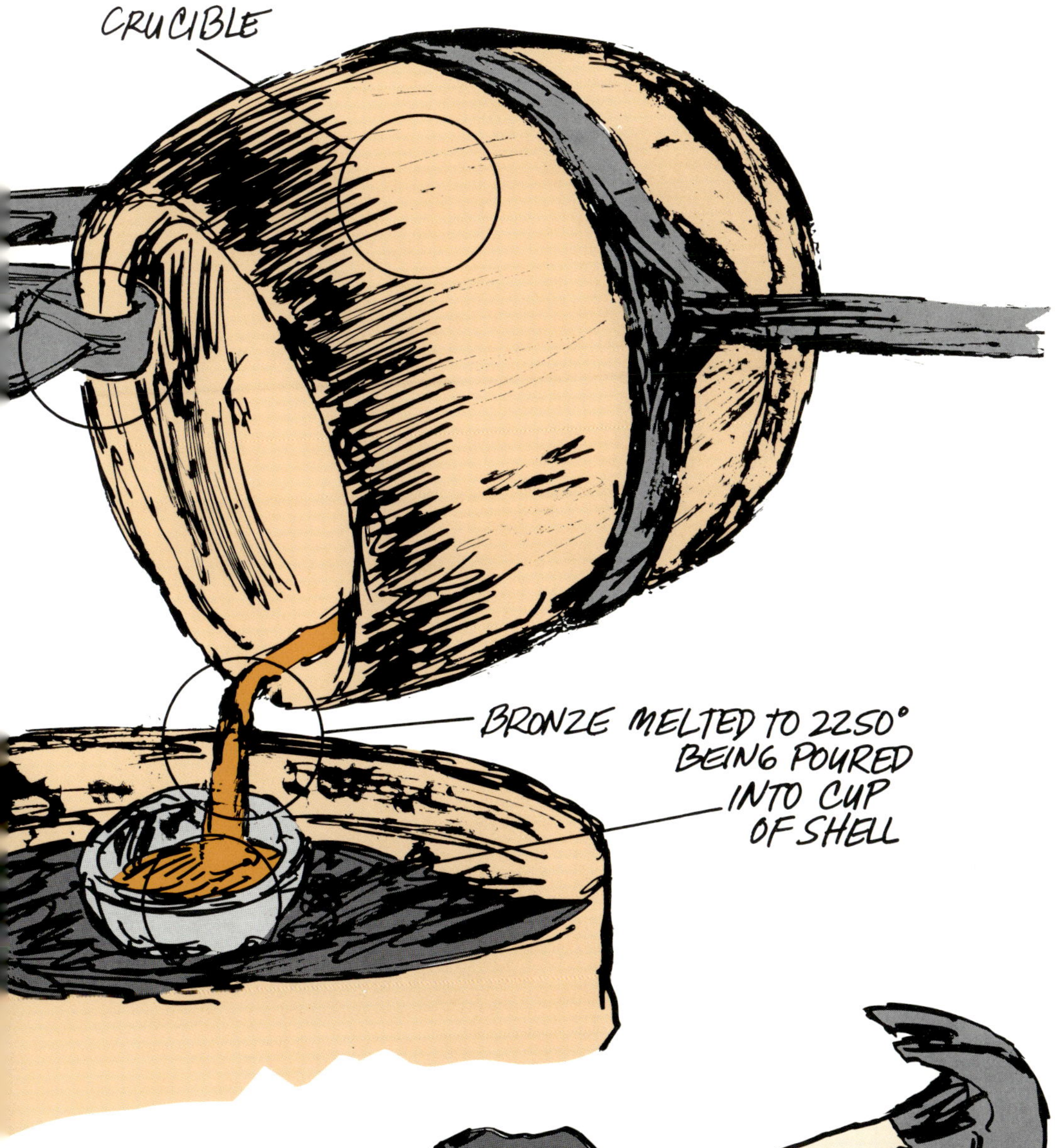

10 Pouring the Molten Metal

The metal, which is heated to approximately 2200° is then poured into the mold.

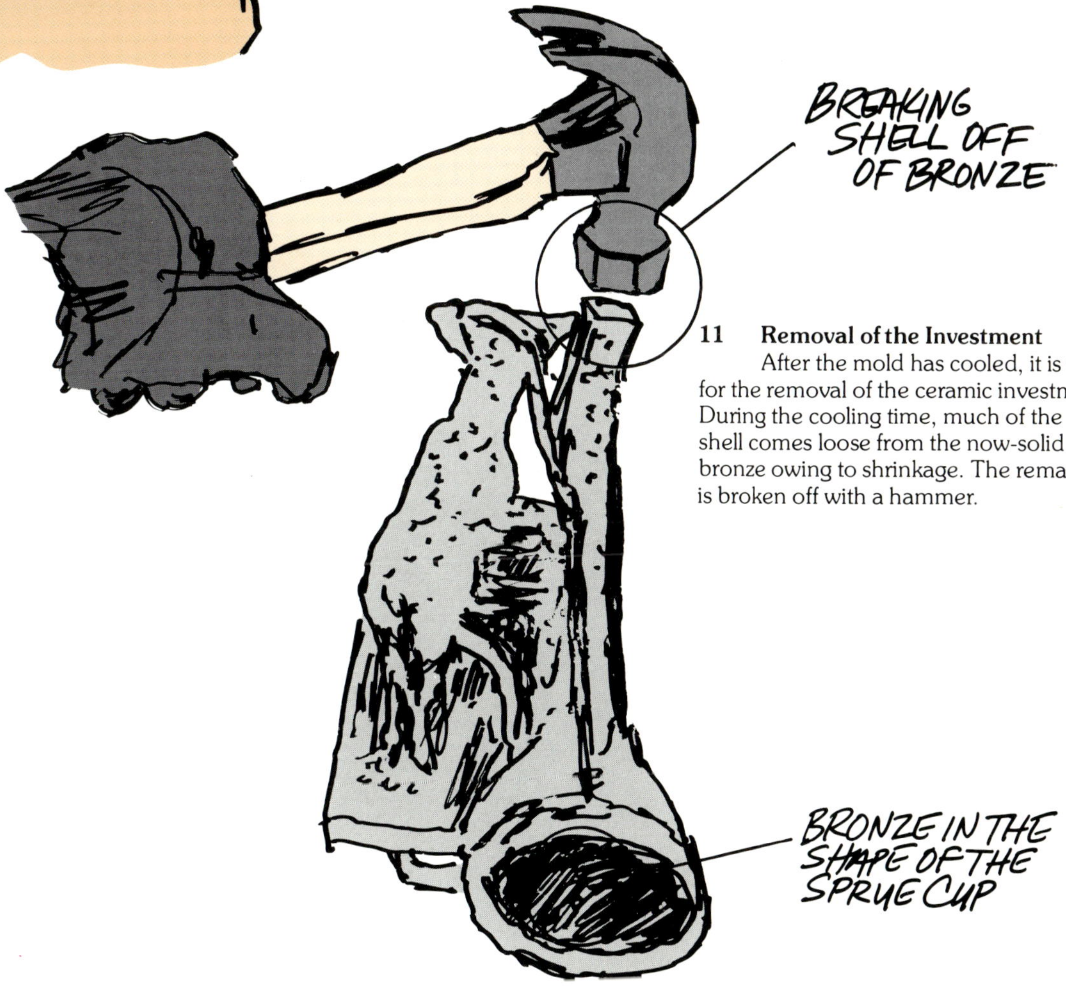

11 Removal of the Investment

After the mold has cooled, it is ready for the removal of the ceramic investment. During the cooling time, much of the shell comes loose from the now-solid bronze owing to shrinkage. The remainder is broken off with a hammer.

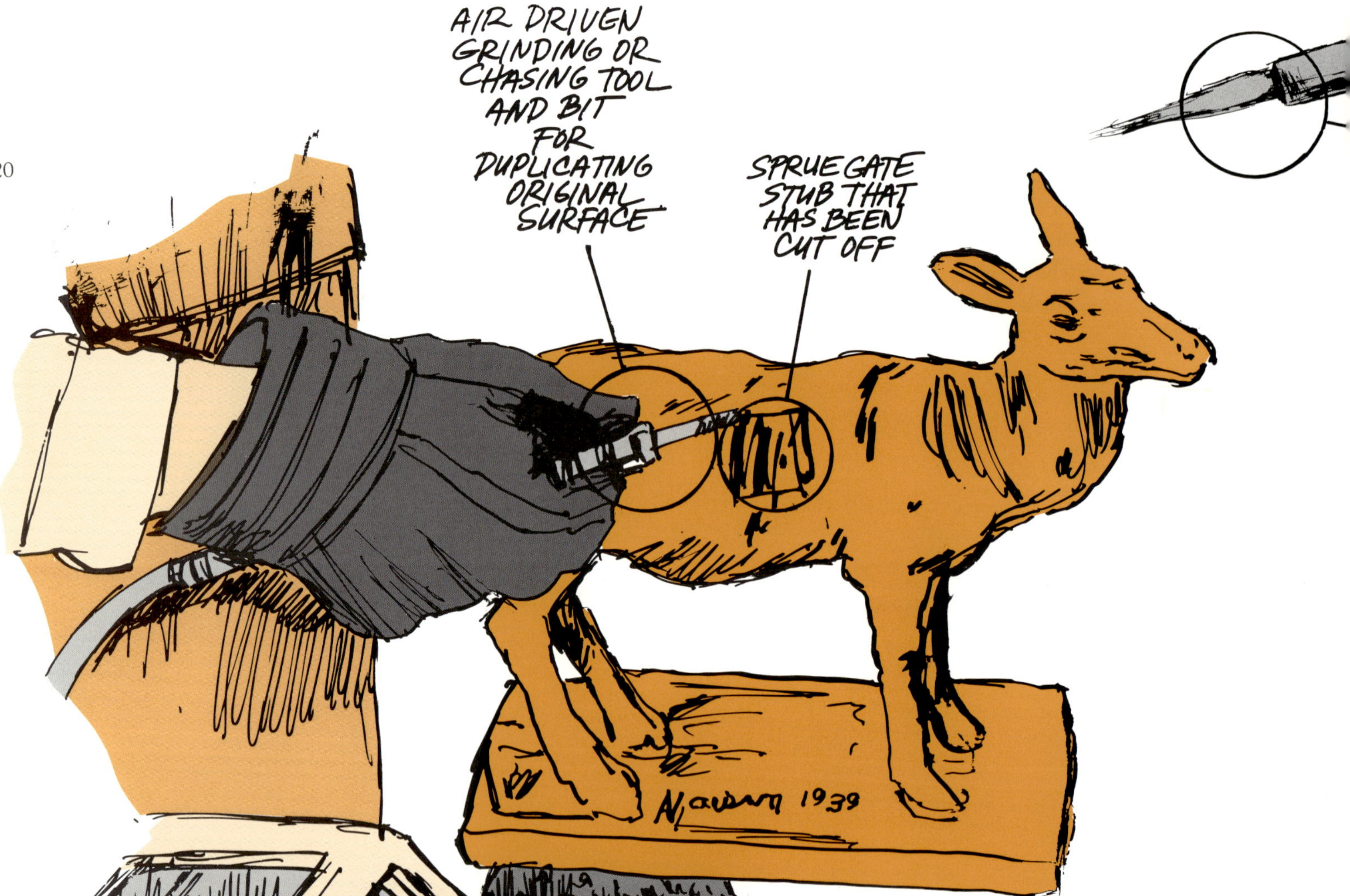

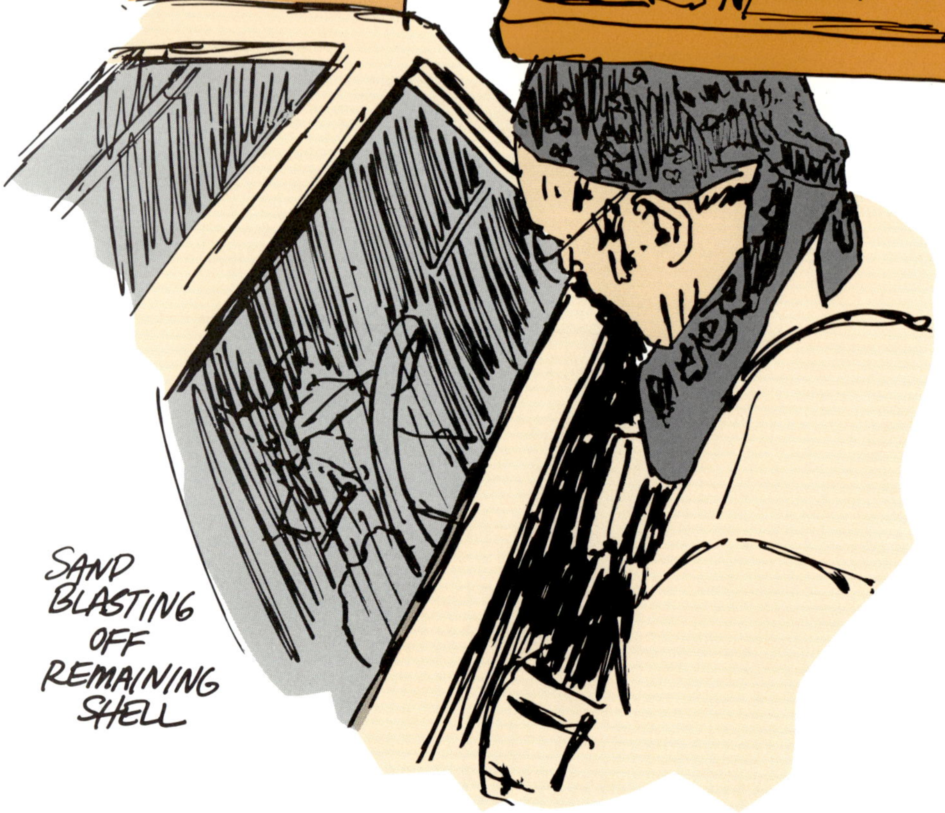

12 Chasing and Sand Blasting

Next, the piece that was removed from the body of the wax mold for the acceptance of the ceramic material into the opening of the elk sculpture is welded into place, and the sprues and venting gates are filed off and smoothed. At this point, the bronze is sandblasted, and the many hours of chasing the bronze begin. This is an important step, and the expertise of the craftsman establishes the final sculpture, true to the original mold.

With a solid investment which some foundries still use, especially in Europe, venting gates are necessary with the ceramic shell process. The permeability of the shell allows the craftsman to sprue without vents.

13 Patina

The last step involves the application of the patina. The piece is placed in a heat chamber for preparation. When sufficiently heated to open the pores of the metal, the various types of acids and chemicals are applied with a torch—different types with varied amounts of heat produce the desired shades and effects.

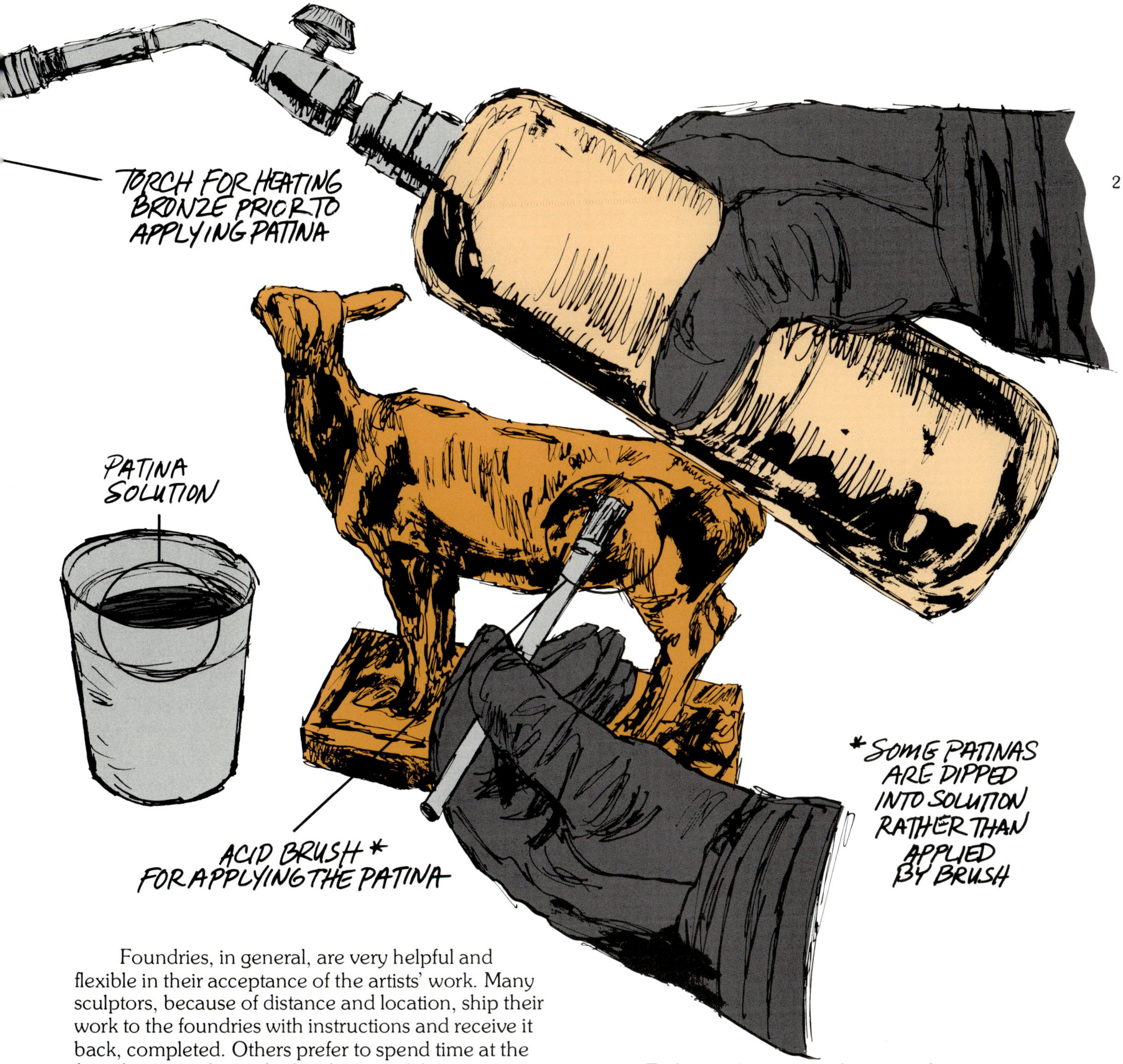

Foundries, in general, are very helpful and flexible in their acceptance of the artists' work. Many sculptors, because of distance and location, ship their work to the foundries with instructions and receive it back, completed. Others prefer to spend time at the foundries in order to do the chasing of the wax; some, the chasing of the bronze; and many elect to apply the patina, but whatever the desire, it takes many hands to master all the complicated steps involved.

As in many professions, there is no one set way to complete a "bronze." This art, handed down from "Father to Son," has many variations. In the past, it took days to build a hot fire in a brick oven to accomplish the burn-out or "Lost Wax." Today, with electric furnaces and high-speed blowers, the process is greatly speeded up. In the past, much emphasis was put on the gating system and venting for the release of the gas build-up.

Today, with newer and improved ceramic investments, the gas build-up is lessened by migration through the more porous shell. New, high-speed grinding tools and chemicals also add to the efficiency and beauty of the work.

I hope that a better understanding of the Art Form, "Sculpture to Bronze," will result in a greater appreciation of the talent of the creator and the expertise of the craftsman.

"Sculpture to Bronze" has many variations. With this new and advanced technology, much of the process has been improved, which adds much to the quality and diversity of the finished work.

Adrien Alexandre Voisin
Artist and Sculptor

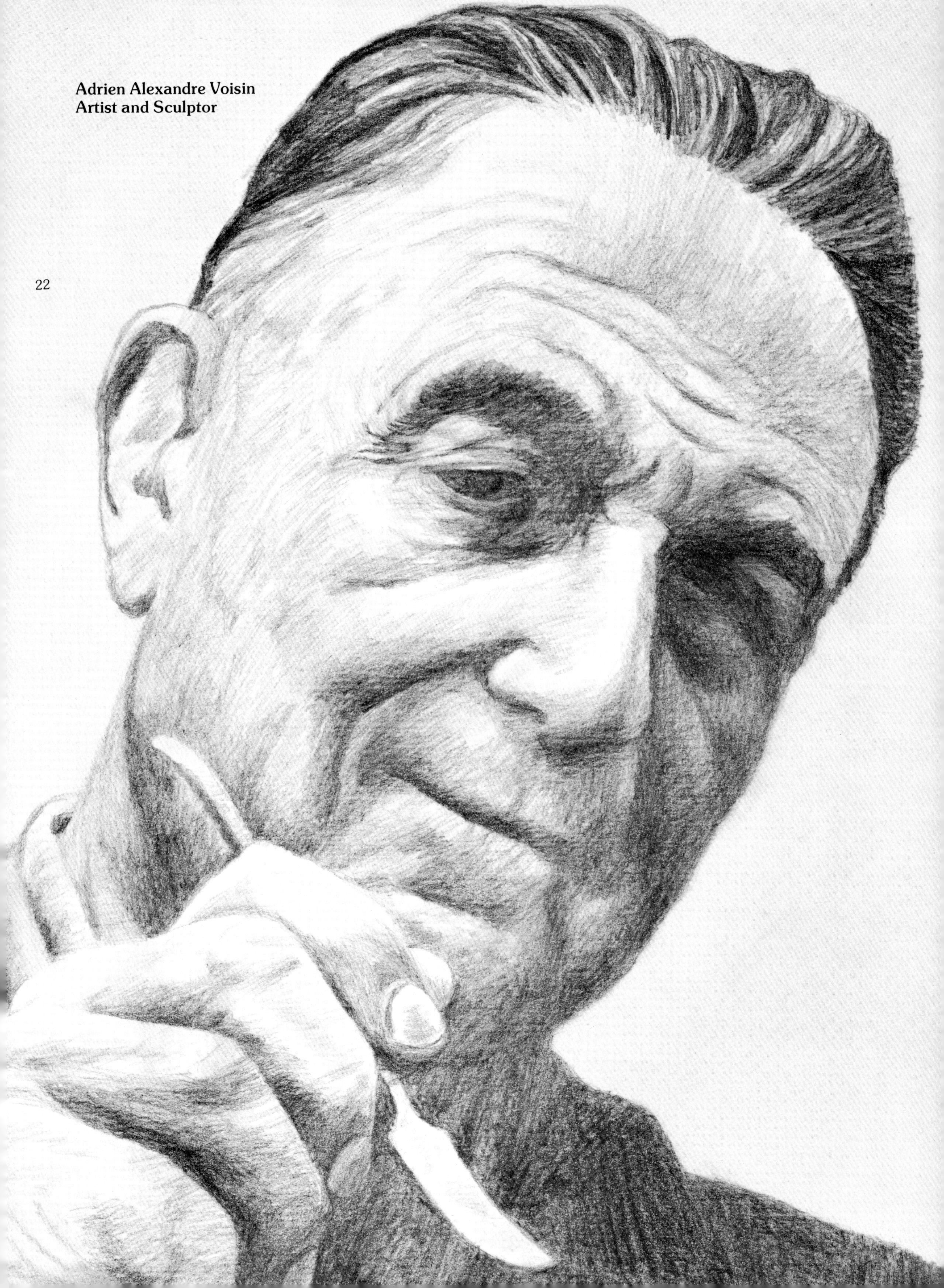

ADRIEN VOISIN

BIOGRAPHY

A noted art critic once said, the four greatest sculptures of the American Indian are Cyrus Dallin, Phimstor Proctor, Earl Fraser, and Adrien Voisin. Voisin will, most certainly, go down in history as the artist-recorder who succeeded in capturing the American Indian in sculpture as Edward Curtis has done in photography.

Adrien's father and mother were both from France. His father, Alexandre, had been a French officer during the Franco-Prussian War, and had come to San Francisco from France in 1872.

It was there, because of his knowledge of horses, he became acquainted with Colonel John B. Stetson, the father of Stetson Hats and shoes. Adrien's mother also was living in the San Francisco area. She was staying with her brother and had opened a school in the old Palace Hotel where she taught French. A short time after their marriage they drove with Colonel Stetson from San Francisco to New York by horse and carriage. This was in 1873. Later, they moved to Islip, Long Island, New York, where Adrien was born, October 4, 1890. Their only other child was Adrien's sister, Marguerite, and she was born two years later.

Training horses was Alexandre's profession, and better opportunities opened for him at Newport, R.I. where he established a riding academy at the foot of Morton Avenue. Here he instructed America's "400" in the equestrian arts.

Beginning at the age of seven, unlimited opportunities presented themselves to Adrien. Newport was in its prime, celebrated not only for its wealth and high society, but also for its charming soirees of intellectuals. It served as a cultural center for the eastern U.S. During his youth, Adrien became acquainted with many people that helped change and influence his life. He had frequent conversations with people such as John Lafarge, Charles Cushing, and Henry Clews, son of the famous financier, marine painter Elijah Boxter, and Sargent Kendall of the Yale School of Fine Arts.

Adrien's first ambition was to become a naturalist. Lacking interest in the customary course of studies, he dropped out of school and attached himself to a pioneering taxidermist, who was willing to put up with the youth for a very small sum. There, Adrien and his master learned to stuff animals, and together, improvised forms out of wire and glass eyes and introduced them to the public for the first time. For eight years, he tramped the woods for subjects, gradually assembling a large collection of New England wildlife, that he arranged in the attic of one of his father's stables. It became his own museum. With these years of experience in this form of art, he now became attracted to others. He studied under an old French master of woodcarving who agreed to take him on as an apprentice, and for four more years Adrien worked at this art without a salary. When he completed this apprenticeship, he was given a paper stating that he was a master carver — this was his wages for four years of hard work.

During this time, Adrien had been painting part time under the tutelage of Sargent Kendall, and it was Kendall who influenced Adrien's decision to go to Paris for further study. With little money, he sailed to France in 1912, and there enrolled in the Academé Colorossi under Paul Bartlett.

His progress was swift, and from there he moved to the Ecole National des Arts Decoratifs and studied under Camile Lemaire — and then to the greatest of them all, the Ecole des Beaux Arts where he worked with Mercer and Injalbert. It was here that he came to realize that sculpture, not painting, was his field. Encouraged by his teachers, he modeled Indians. Years before when Adrien was a boy, Buffalo Bill's show had left a deep impression on him, and the idle sketches he made of Indians at the Beaux Arts attracted the attention of his master. "Why don't you make Indians?" the old Frenchman suggested, "It's never been done before."

Voisin followed this advice, and instead of merely repeating, with variations, classical and mythological subjects exploited by Rodin, Bourdelle, and their host of imitators, he set about modeling the Indians of his own country.

During Voisin's studies in Paris, World War I broke out, and Voisin enlisted in the American Ambulance Corps, and later served with the 49th and the 93rd U.S. Infantry Divisions. He participated in the Battle of Chateau Thierry and was seriously afflicted by mustard gas in the Toul sector. After returning to the United States, he was discharged at Fort Leavenworth, Kansas, in 1919.

Leaving there, Adrien decided to go to California because of his health. An important commission during this period was his appointment as head sculptor at the San Simeon estate of William Randolph Hearst. As such, he coordinated the European art collections and added the sculptures and frescoes which still may be seen at the Hearst castle.

His commitments then took him to Spokane, Washington to mold the decorative designs for the Paulsen Building. It was at this time that he became acquainted with Vachel Lindsay, a well-known poet and writer. Being many years younger than Voisin, and leading a very Bohemian life also, they became very good friends, and through their travels, met many of the Indians of that part of the country. It was also through Lindsay that Adrien met the Indian, John Clarke, whose other name was CUT-A-PUIS, who he encouraged to become a sculptor. John Clarke was a deaf mute from birth, and Adrien spent many hours teaching him how to sculpt.

Mrs. Voisin was the daughter of an Indian agent, and acted as an interpreter when they went to Browning, Montana in 1929 where they rented an old Bureau of Indian Affairs building and set up a studio. It was here, through a friendship built up with NINA-STUKO, the famous mountain chief, and Two Guns White Calf, that Adrien was able to influence many of these native Americans to pose for his sculptures. It was also at this time, because of their understanding of, and friendship to the Indians that Voisin and his wife Frances were adopted into the tribe in ceremonies, which he described in vivid detail in Robert Harr Kimbro's manuscript. They bestowed on him the name of

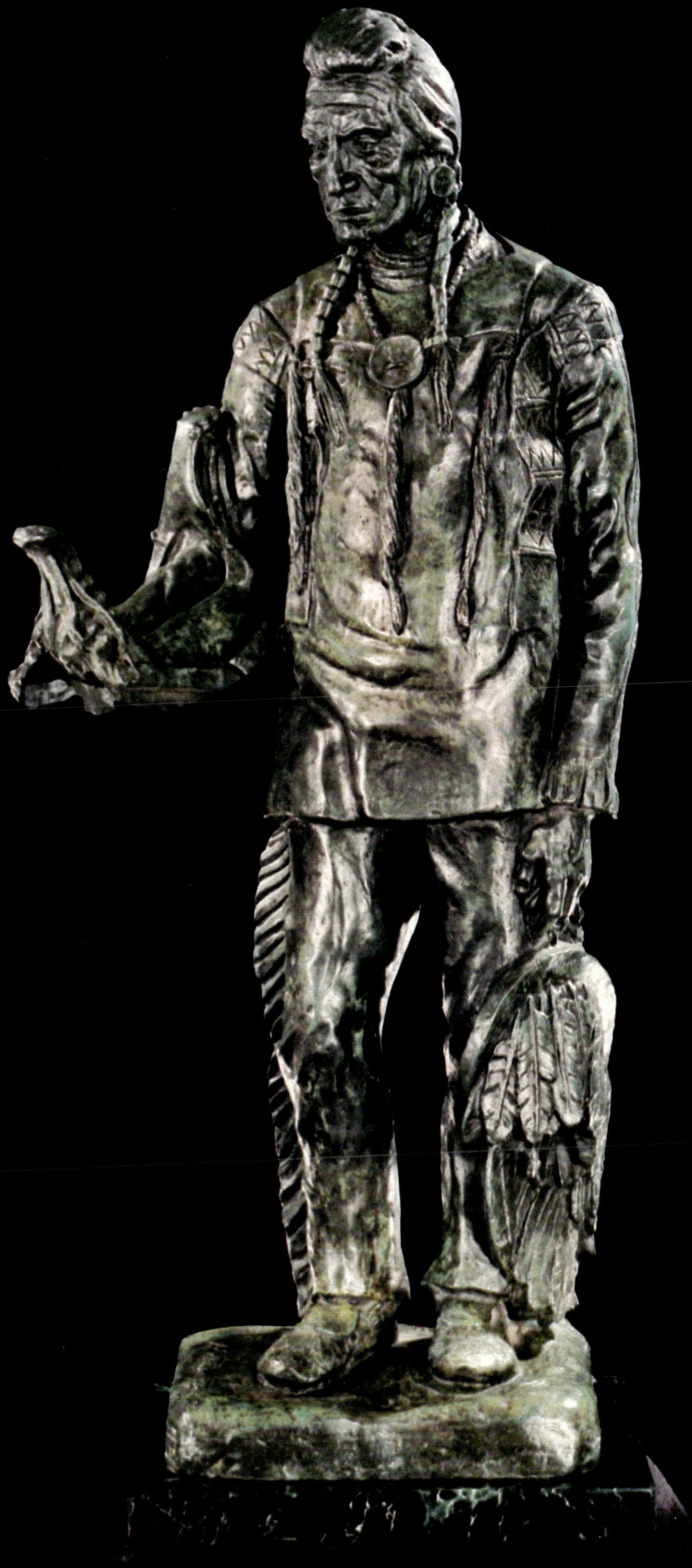

AH-GAW-CHE-NENA, which meant "classic sculptor of the Blackfeet."

In 1929 the Voisins left Browning with the cherished results of his work in sculpturing the Indians he had known. They traveled to New York City, where he took his original twelve clay sculptures to the Roman Bronze Works. It did not take him long to come to the conclusion that because of limited finances, he would find it far less expensive to have his bronzes cast in France. (At that time, the rate of exchange made this possible.) Immediately upon his arrival in Paris, he had his twelve pieces cast before his money ran out.

Then, in March 1930, he had a formal showing at the Galleria Bernheim in Paris. Hailed by prestigious critics as a great artist, Voisin opened a studio there. The following year, the U.S. Government, in an attempt to enter an exhibit from America in Paris, had its shipment lost en route. Voisin was then prevailed upon to represent the U.S. Government in exhibiting some of his work, and in 1931, the twelve Indian sculptures in this book listed 1 through 12, were entered at the Exhibit Coloniale. The opening of this international show received world-wide acclaim and was attended by dignitaries from around the globe. It was at this show that the French government awarded Adrien Voisin the Diplome d'Honeur for the exhibit of his famous twelve American Indian sculptures.

The following year, he entered the International Art Exhibit in Paris. His twelve Indian busts won the Gold Medal. With the Diplome d'Honeur, the Gold Medal, and the favorable reviews written by the critics of that nation, Adrien was now on his way to a successful career. In these years in France, he decided to try his hand at painting again. In 1930, he finished a series of 22 paintings in Toledo, Spain, painting landmarks and buildings. His selection of subjects showed no inkling, however, of the devastation that war was to bring. His pictures have since become a rare record of such places which were destroyed by civil warfare. Some of the famous places which Voisin's brush converted onto canvas were pre-war scenes of Segovia; Fuenterrabia, on the seacoast near Irun; Toledo on the Tagus River; and Granada, all of which had been reduced to shambles during the war.

In 1934 Adrien returned to the United States and spent much time traveling through Montana, Washington, and Idaho, seeking Indians as models. He completed 39 pieces and also sculptured 29 animals native to that region. On November 22, 1935, he arrived at Pendleton, Oregon for the famous Round-up. Here he spent twelve days carrying out his work, making a permanent record of surviving types of Indians in Oregon and Washington. For those 12 days he lived in a tepee in the Indian village while making models of the natives there. Among those he made models of later to be cast into bronze were Chief Red Hawk of the Cayuse and Red Hawk's niece, Melissa Parr, the Indian queen of the Round-up and a member of the Nez Perce tribe. Others were Sam Morris and Poker Jim, well known Umatillas.

Later, in 1935, Voisin set up a studio in Portland, Oregon, and during the day would spend much time at the Washington Park Zoo, sculpturing the animals in that live habitat. Shortly after, he and his wife moved to San Francisco, where they acquired the ruins of the old Albion Brewery near the Navy shipyards at Hunter's Point. It had been built by English masons in 1870. Called a "castle," it was a faithful replica of a Norman cape scene.

For twenty years he labored at its restoration, doing all the work himself, down to the last hinge of its massive doors and the huge andirons in the oversized fireplace. The reconstructed castle became Adrien's home and his studio. He was quite happy there until Frances died after a lingering illness.

Adrien's now lonely life was brightened when he met an old sweetheart, a friend of his earlier years. She was Mary Stearns, a widow, and when she consented to marry him, the two lived for a time in the castle, until he became worried about the change in the neighborhood and atmosphere at Hunter's Point, insofar as they and his valuable busts were concerned.

Eventually, some of the pieces were shipped to the Museum of Native American Culture at Spokane, where they are still on permanent display.

On August 22, 1978, Adrien and Mary arrived at Spokane as honorary guests of the museum, to view the display of his work. They were met at the airport by Father Schoenberg, director of the museum, and during their three day visit Adrien was presented with a special gift.

It was the YAKIMA DANCER, which Adrien had sculptured in clay in 1936 and was never cast. And so, he saw it for the first time in bronze. This gift to him was a paradox unequaled in Adrien's lifetime.

The other parts of his vast collection, including the slates that Adrien had carved, were taken to Los Angeles, California, where he had many of them stored at his sister Marguerite's home.

Adrien's move to Southern California, with Mary, would no doubt have been the happiest part of his life except for the illnesses that plagued them both. After eighty-eight years of eventful and creative living, Adrien Voisin went to his Happy Hunting Ground on May 7, 1979.

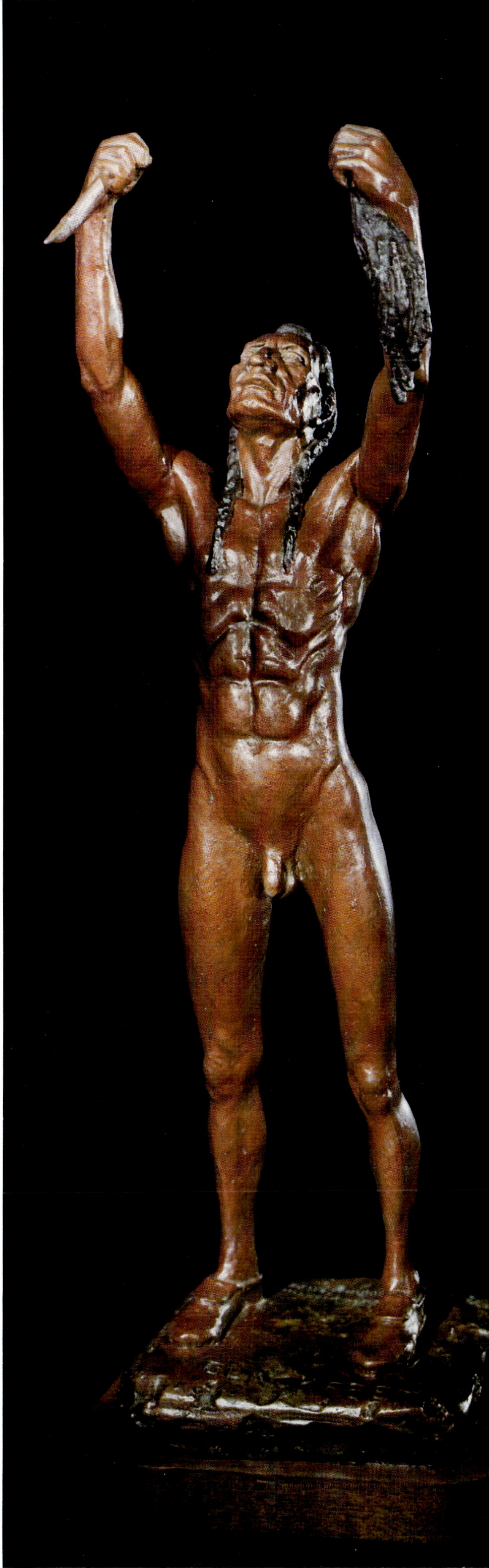

SCALP PRIEST

In his moments of simple acceptance of things as they were, he was conscious of his own immortality as a sculptor, and he was one of the greatest; but it was his tragic fate that twice his career was cut short by world events, namely, World War I and World War II. He lived for three decades in relative obscurity, after having received praise from all over the world.

Adrien Voisin was a very humble and modest man. Our conversations always drifted from himself to his opinions and theories on art. During one of the last interviews that I had with him, he dwelled on his favorite subject. I asked him what his feelings were about the differences between sculpturing and painting. "Sculpturing," he said, "is the primal mode of artistic expression, appreciated and understood by men of all races, cultures, and ethnics because it makes its appeal to the mind through the strongest of senses — touch. Sculpture is the greatest of mediums because it not only requires three-dimensional treatment, but it also has the ability to express with force many things that cannot be said with words. The sense of touch is both the oldest and the youngest of the major senses, as we experience them. As infants, we get our first impressions of feeling created by the sense of touch, and later, we learn to translate these feelings to the eyes by developing associations between the reactions of the same object to touch and sight. In old age when our vision is impaired, we resort to the sense of touch again. This perhaps explains partly the universal appeal of statuary, regardless of whether we run our fingers over the graven surfaces or feel through our eyes. The sculptor must have many visions. He must see from all sides. From the top, he must study his subject from all angles. While the painter paints one profile, the sculptor must do a thousand and have them so blended as to produce a work that is harmonious in every detail. Sculpture falls into two major classifications, each of which presents intriguing problems. There are portraitures of human beings as well as animals in the allegorical in which the artist is free to express his own thoughts and emotions. "I believe," he said, "there is a certain continuity in the minds of men that has no form of expression in words or in any other medium, other than sculpture. Every man has a basic common feeling when confronting some great mountain, glacier, ocean, or natural upheaval — a mysterious something that words cannot reach and yet can be caught and told in sculpture. Sculpture is the unspoken word that connects the past with the present and which will carry the message of the present far into the future after all other forms of expression have vanished."

Voisin gave the following advice: "give up everything else and become an expert craftsman, well founded in the precepts of the past; then carry the torch a little further on..."

His highly talented hand helped chisel and mold new dimensions in appreciation of the art for those who came in contact with him. The collecting of this series of his monumental works so superbly crafted and being able to put into writing this manuscript have a personal satisfaction to me. I only wish that Adrien Voisin was still here to see this gathering of his accomplishments and efforts which can now be enjoyed and seen by art enthusiasts in museums around America.

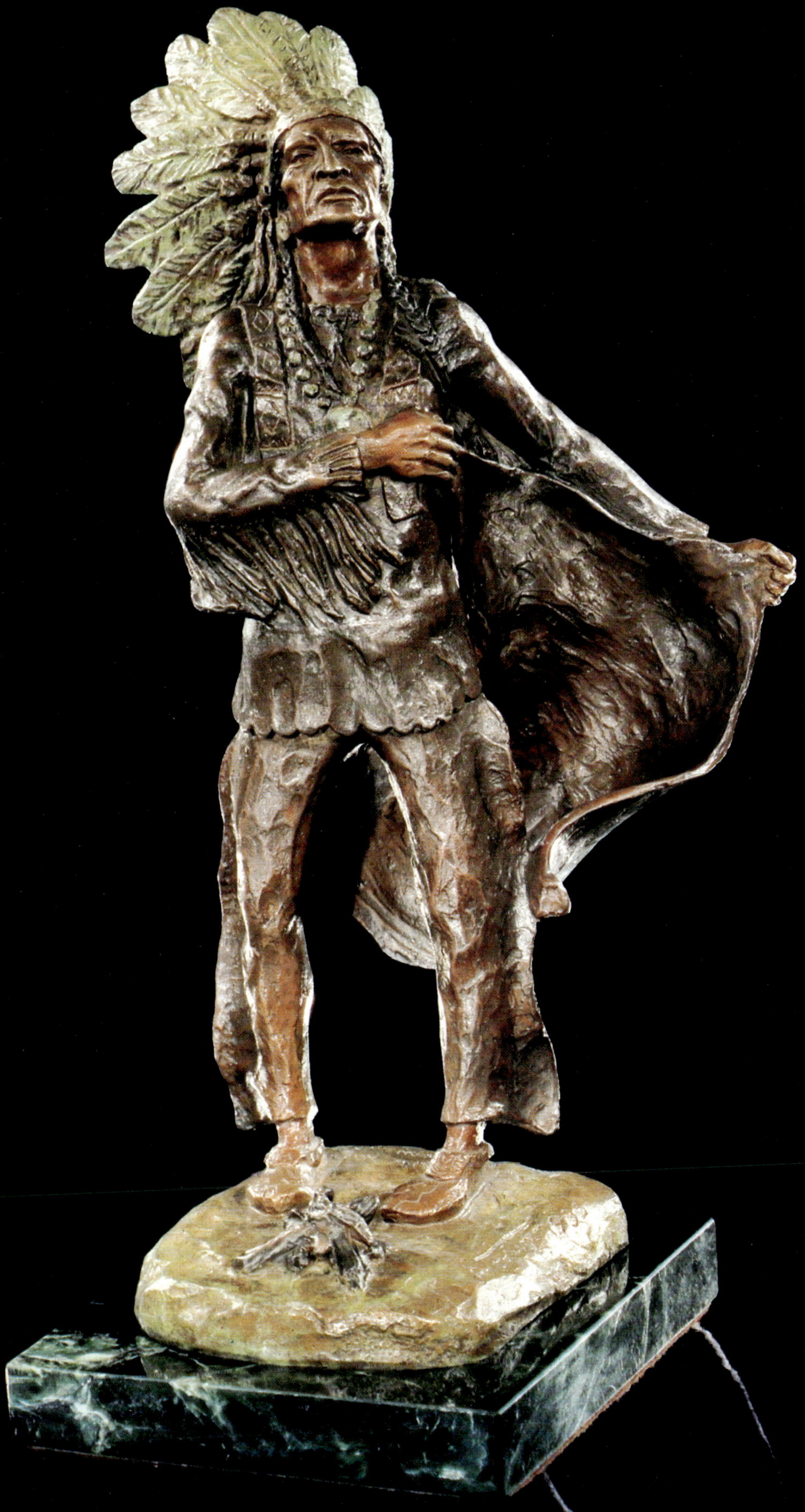

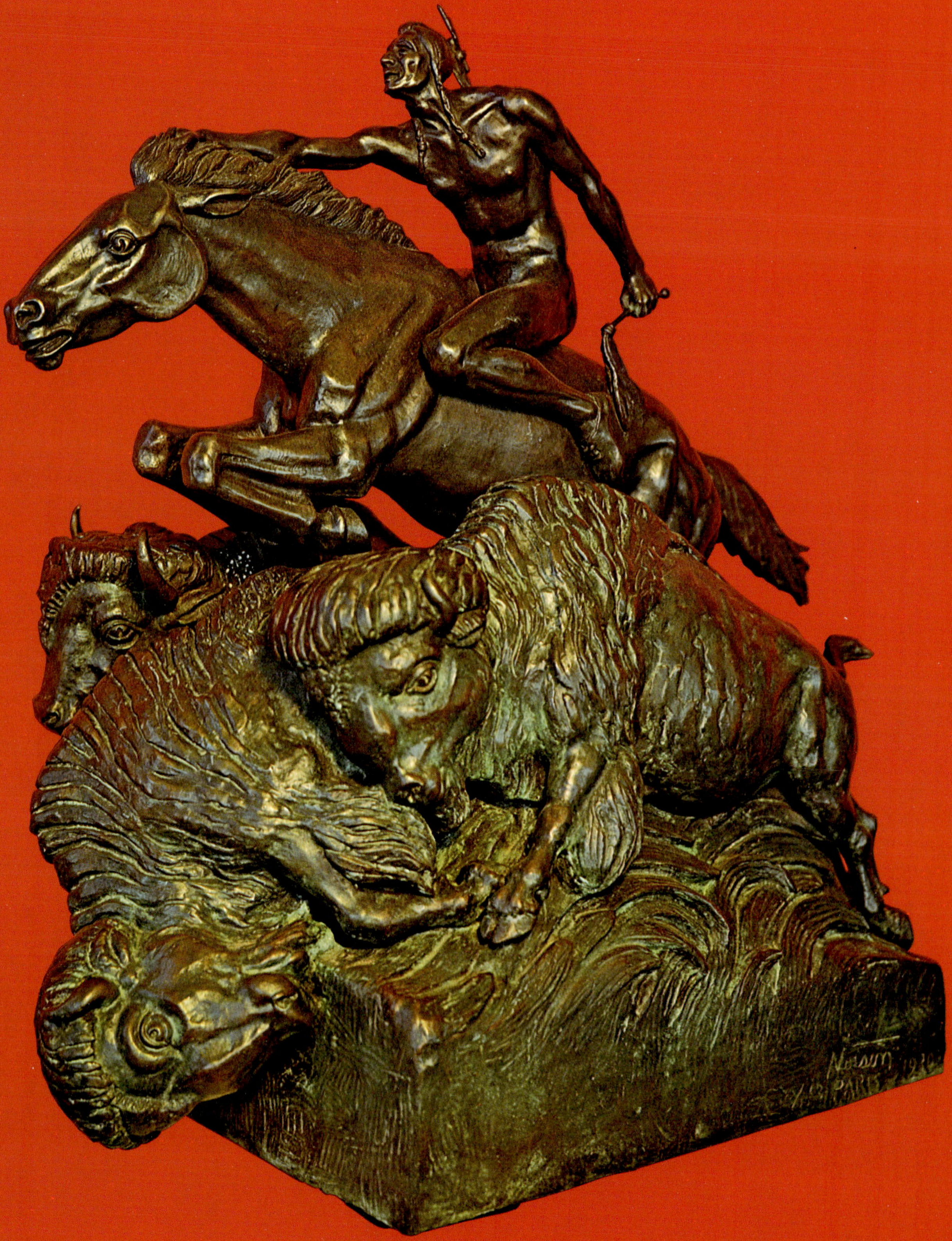

PARIS SHOWS

HONOR FOR VOISIN IN PARIS ART SHOWS

In 1929, Voisin with his twelve plaster sculptures of the Indians he had just completed in Montana, returned to France where he had them cast into bronze at the A. Salsuani Foundry in Paris. In 1930 he held a formal show at the Galleries Bernheim-Jeune in Paris.

Hailed by the critics as a great artist, he decided to open a studio in Paris. In the following year his Indian collection was loaned to the United States Government for its exhibition at the Exposition Coloniale in Paris. For this the French Government awarded him the Diplôme d' Honneur. Voisin also received a gold medal in the International Art exhibition in Paris in 1932.

1 **HAPPY HUNTING GROUND.** Buffaloes and Hunter—Blackfeet. Bronze Height 26 inches. Dated 1930.
This monumental work carries the imagery of the American Indian to its finest expression in sculpture form. During a conversation while sculpturing "Nina-Stuko" Mountain Chief, Voisin asked him, "What are your thoughts of the future world?". He answered, "I want to go to the Happy Hunting Ground and ride a great stallion over a herd of charging buffaloes." This was Voisin's inspiration for doing the large bronze, "The Happy Hunting Ground," and a tribute to a great friend and honorable Indian Chief.

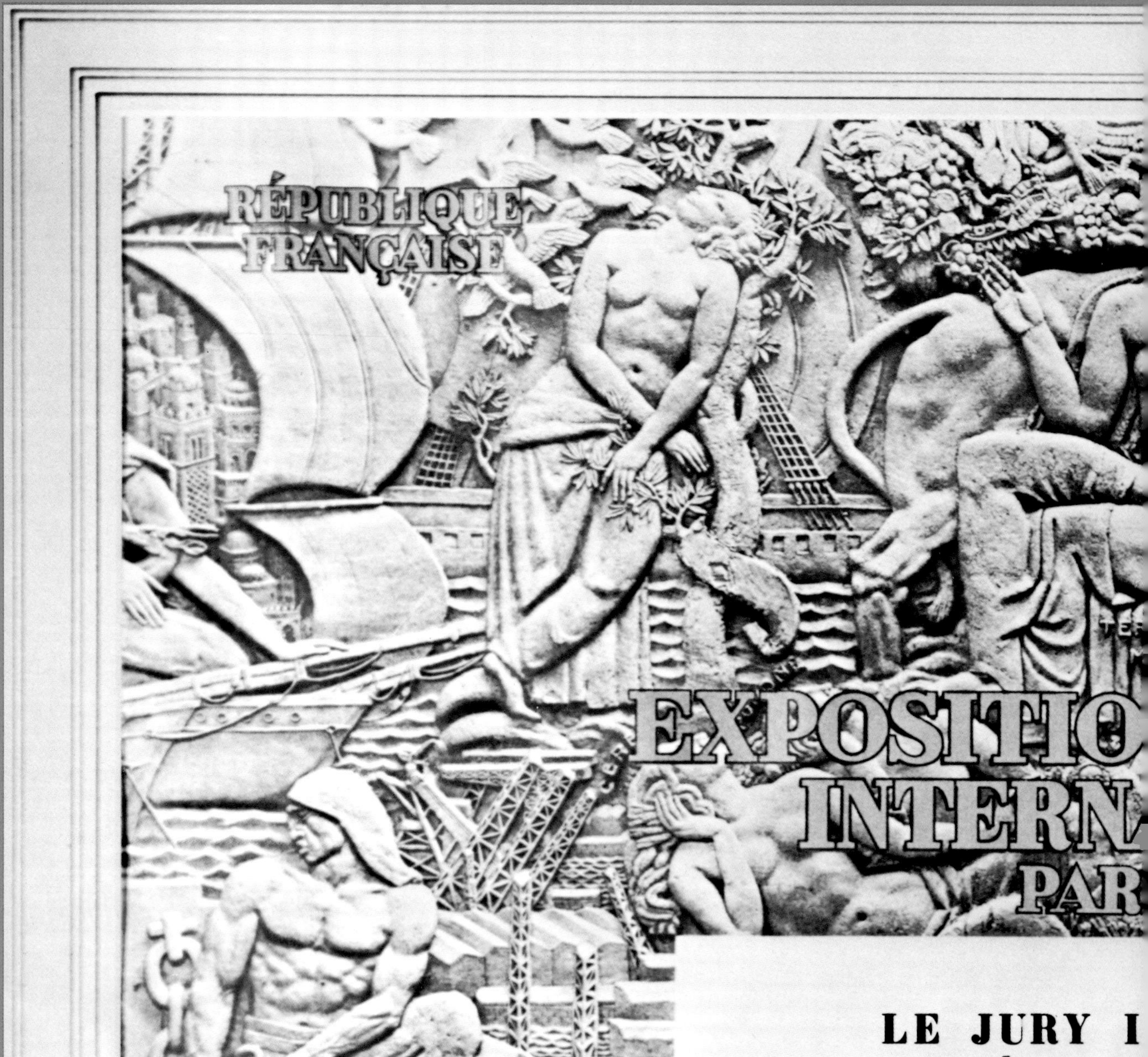
RÉPUBLIQUE
FRANÇAISE
EXPOSITIO
INTERN
PAR
LE JURY I
DES RÉCOMP
un Diplô
à Monsieur A
Groupe XV
Le Commissaire général de l'Exposition :
Le Commissaire de la Section :

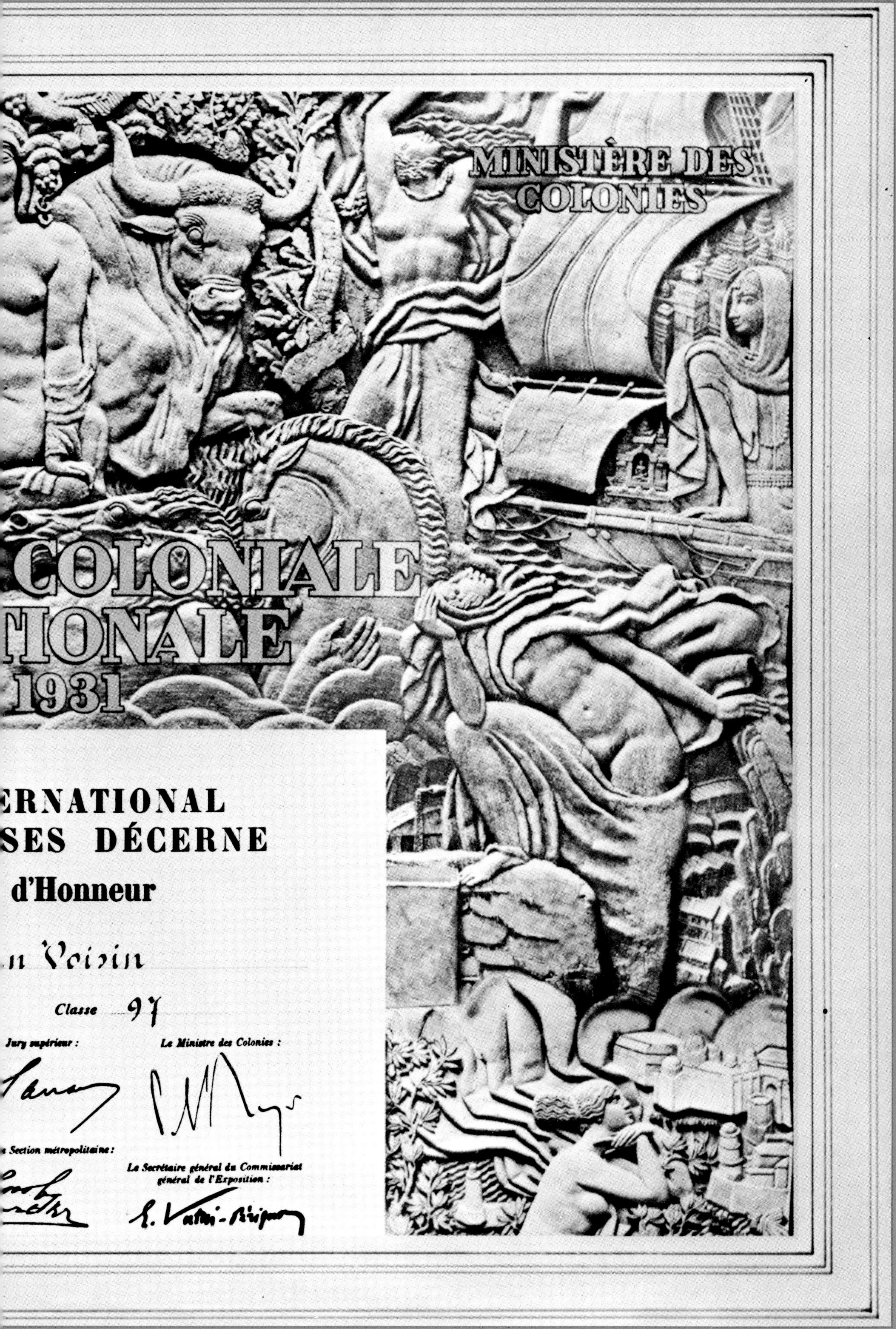
MINISTÈRE DES COLONIES

COLONIALE
TIONALE
1931

ERNATIONAL
SES DÉCERNE
d'Honneur

n Voisin

Classe 97

Jury supérieur :

Le Ministre des Colonies :

Section métropolitaine :

Le Secrétaire général du Commissariat général de l'Exposition :

1931— Adrien Alexandre Voisin Awarded the Diplôme d'Honeur by the French government

Published News Release
ADRIEN VOISIN, AMERICAN SCULPTOR
by
B. J. Kospoth

From: ARGUS de la PRESS
"Voit Tout"
37, Rue Bergere
Paris, France
23 Mars 1930

Among the innumerable one-man shows staged by American and other foreign artists in Paris this season, Adrien Voisin's exposition at the Galeries Bernheim-Jeune has compelled the attention of French critics and amateurs of art to an exceptional degree. Mr. Voisin owes this remarkable success to the fact that he is one of the rare American artists who draw their inspiration from purely American sources. Whenever an American has the courage to present American subjects to the Parisian public, he is certain to stir its appreciative interest.

Adrien Voisin is a sculptor of the American Indian. He has specialized in recording in bronze the keen, rugged features of a great, mysterious, and dying race. Frenchmen, who have tired of seeing American artists repeating over and over again *motifs* already immortalized by the great French masters, have turned with relief to his splendid heads of Blackfeet chiefs and warriors. Following his exposition at Bernheim's, Mr. Voisin is planning to show his work at the coming Paris "Salon," which opens next month where it will undoubtedly likewise excite much curiosity and admiration. For, apart from the romantic interest which the American Indian has for all Frenchmen, Mr. Voisin's work has qualities of strength and simplicity that place it in the front rank of modern sculpture.

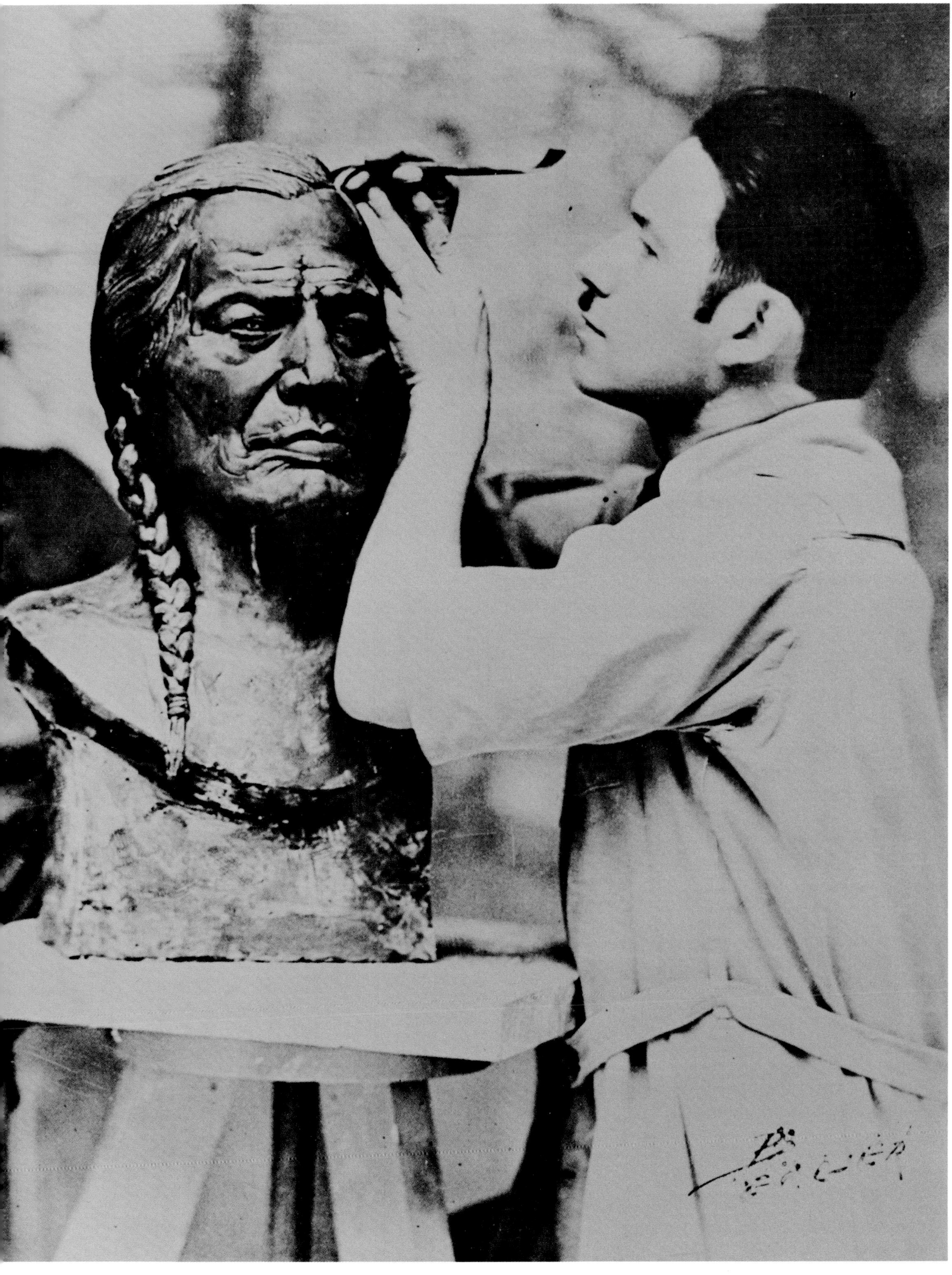

Adrien Voisin in 1929—A Legend Within His Own Time

EXOTIC ART

We are acquainted with the fine studies written by Gilbert Chinard on American exoticism in French literature during the last four centuries. These studies were a revelation for many people. Although we had a general idea of the benefits derived by Marmontel, Prévost, and especially Chateaubriand, from primitive nature and human types foreign to our civilization, we did not know perhaps that our older authors were also thus inspired. It was Chinard's merit to show, by citing French authors often unknown for this aspect of their work, all the resources to be found in the virgin landscapes over there and in the whole American Indian culture.

Adrien Voisin saw that this domain could very well furnish him with subjects for his art. Happy idea! Doubtless preceding him there were painters and sculptors who occasinally chose themes from this unexplored field. In the United States, Cyrus Dallin, to cite only one, accomplished something very beautiful in this genre with his "Appeal to the Great Spirit." One cannot praise too highly the sincerity, the fervor and the sheer artistry of this work. But no sculptor had yet, I believe, specialized in this branch of art. French artists had the plausible excuse of distance. Where would they go to look for their models? American sculptors could say almost the same thing. For one can travel a long time all across the United States without meeting a single authentic descendant of the great races which formerly peopled the land. However, some still exist. One must know where to find them.

Adrien Voisin had this good fortune. His practiced eye immediately seized the picturesque qualities of these individuals who had remained so close to nature. These heads, so strange and beautiful, opened for him perspective on an unknown world. He lived with the Indians, became their friend. The more he observed them, the more clearly he saw what these faces could offer in transposition into sculpture, what richness of effect they presented, what new elements they brought to art. It was in a way the unpublished. For no one had yet fully exploited this field.

He applied himself then, working not for style but according to life, conscientiously, scrutinizing these masks in order to reveal their secret, trying to express their unique character, and to depict the basic characteristics by which they are distinguished from us. In fact, it was not a question of depicting Indians of fantasy, suited for the stage set of a comic opera, but true sons of the woods, whose image he had before his eyes, an image which was being erased and which soon will be only a memory.

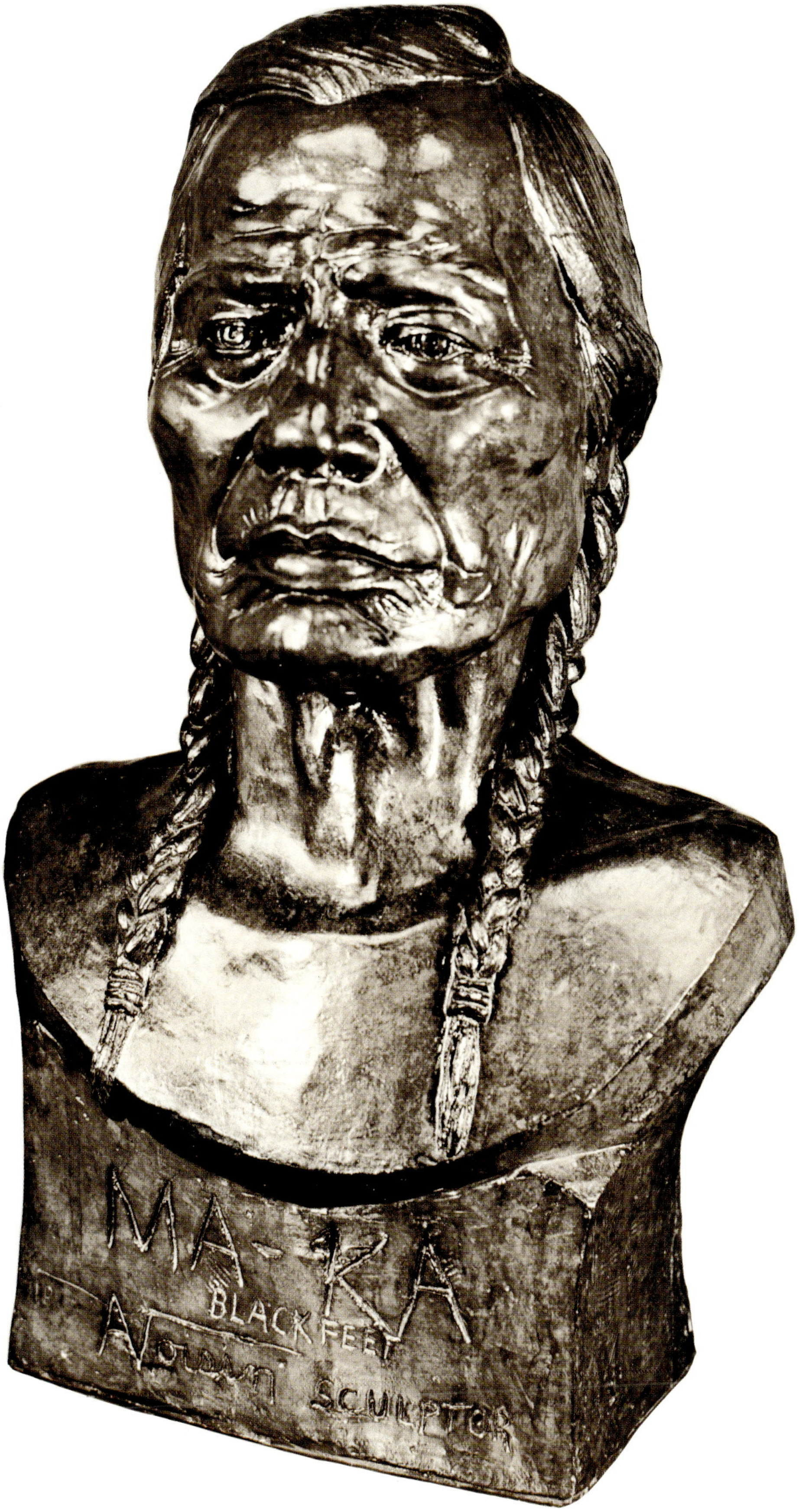

2 **CHIEF MA-KA.** Mike Short Man, Blackfeet—Height 21 inches—Dated 1929. An expert in the Indian sign language, and considered the best in North America. He interpreted for Voisin on his first visit to Browning, Montana. MA-KA married Louise, known as FINE MASSACRE. She was the daughter of Wolf Man and Snake Woman. "I would much rather do Indians," Mr. Voisin said, "Their faces express so much more character than the White man's." After seeing the bust—critics are inclined to agree.

The exposition, to which the Parisian public is invited, is the fruit of his works in this direction. It is not necessary to emphasize the documentary interest. The exposition offers students of ethnology a vast field of observation. Their scientific curiosity will adapt easily to the contemplation of the last representatives of a dying race. As to the artistic value of the exposition, I would prefer to leave the task of pronouncing on it to the connoisseurs and men of the art world. One does not expect a lay person to put forth a judgement on the merit, technique and esthetic of such a collection. At the same time, I am not going too far in predicting a great success of this exposition. The works of a Frenchman from overseas who has applied to exotic subjects the principles of art and the rigorous training received from our masters and in our great schools are sure of being greeted with honor in our country. Their quality will command admiration. These Indian figures are portraits and one, among others, will not fail to attract particular attention; that of Vachel Lindsay. The great poet of the American West is in fact an authentic descendant of the Blackfeet Tribe, as is well attested by his facial characteristics; and this does not prevent him from having genius.

Adrien Voisin gives honor to America and to France—nations to which he is equally indebted. Paris, which reveres talent, will not stint in its admiration of this far-away son, who in the Great War (World War I) came to defend the soil of his forbears and who now, on American shores, continues the tradition of beauty by which France has made itself an immortal name.

Prince Henri d'Arles
Laureate of the French Academy
Translation by Mary Rose
Cunningham

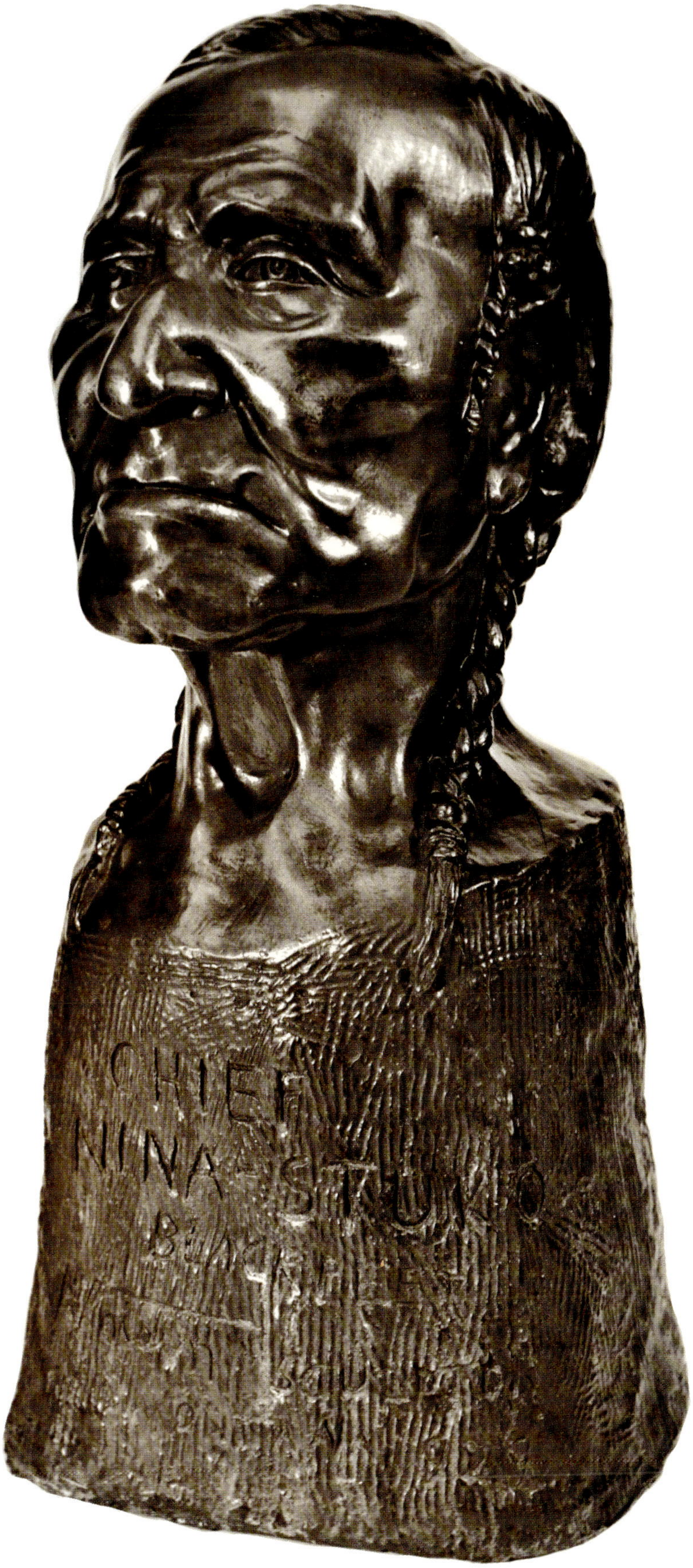

3 **NINA-STUKO.** Mountain Chief. Blackfeet. Bronze Height 21 inches. Dated 1929. He had been the leader and hereditary Chief of the Confederation of the Five Nations, and had fought in 28 battles against the Sioux. He was known and famous for his sagasity and honesty. There were three Mountain Chiefs named NINA-STUKO, each the father to the next. One signed the treaty in 1868 as Piegan Head Chief, though Three Bulls did the negotiating for the entire Blackfeet nation. This is the same chieftain whose band murdered Malcolm Clarke and then escaped to Canada. The last NINA-STUKO Mountain Chief died in 1942 at 103 years old. When Voisin worked on his sculpture of Mountain Chief, whose failing eyesight would not allow him to see the artists finished project, Voisin had him inspect it through the sense of touch. He insisted that Voisin make him with his eyes open. He said, ''me not always blind, me want children to see me with eyes always open.'' Mountain Chief's Granddaughter, Rita Spotted Eagle, is now living in Browning, Montana.

Some of the dignitaries attending the opening of the Exhibition at the Exposition Coloniale in Paris 1931

Monseigneur, the duke and prince of Grenelle, Paris
Mr. Jean Guiffrey, conservator of the Louvre Museum.
Mr. Maurice Denis, "The Priory," Saint-German-en-Laye, Seine-et-Oise.
Mr. Maxim Real del Sarte, sculptor, 6 rue de Vercengetorix, Paris.
Mr. George des Vallieres, painter, Paris.
Mr. Fernand Livoire, editor at the *Intransigeant,* Paris.
Mr. Arséne Alexandre, art critic at *Figaro* (French newspaper).
Mr. Francois Veuillot, 9 rue du Pré-aux-Clercs, Paris, France.
Mr. René Domvrie, 10 rue de Pré-aux-Clercs, Paris.
His Excellency Monsignor Alfred Boudrillart, Rector of the Catholic Institute, rue de Vaugirard, Paris.
His Excellency, the Ambassador of the United States, Paris.
His Excellency, Mr. Phillipe Roy, Minister from Canada.
His Excellency, the Minister from El Salvador, Paris.
Mr. Alexandre Threnaud, in care of Cartier, jeweler, rue de la Paix.
Mr. Lucien Crozet, 20 rue St. Sulpice, Paris.
Mr. Paul Lefebre, bookstore proprietor, rue de Rennes, Paris.
Mr. Charles Maurras, Director of the Action Francaise, Paris.
Mr. Robert Regnier, in care of the Secretariat of the Institute, Paris.
Mr. Edward Champien, bookstore proprietor, quai Voltaire, Paris.
Mr. George Valois, 6 Place du Panthéon, Paris.
Mr. Lucien Romier, Paris.
Mr. Andre Siegfried, Paris. Author and modern playwright. Very well known in France.
Mr. W. Morton Fullerton. American journalist and literary critic living and working in France for a large part of his life.
Mr. Francis Coty. Director of Figaro in Paris. Figaro is a long time and well known newspaper, strong on literary, artistic and dramatic criticism—no longer published.
Mr. Robert de la Sizeronne. Art Critic in Paris. Very well known in periodical, literary, political and cultural articles.
Mr. Maurice Alary, 24 rue Cassette, Paris.

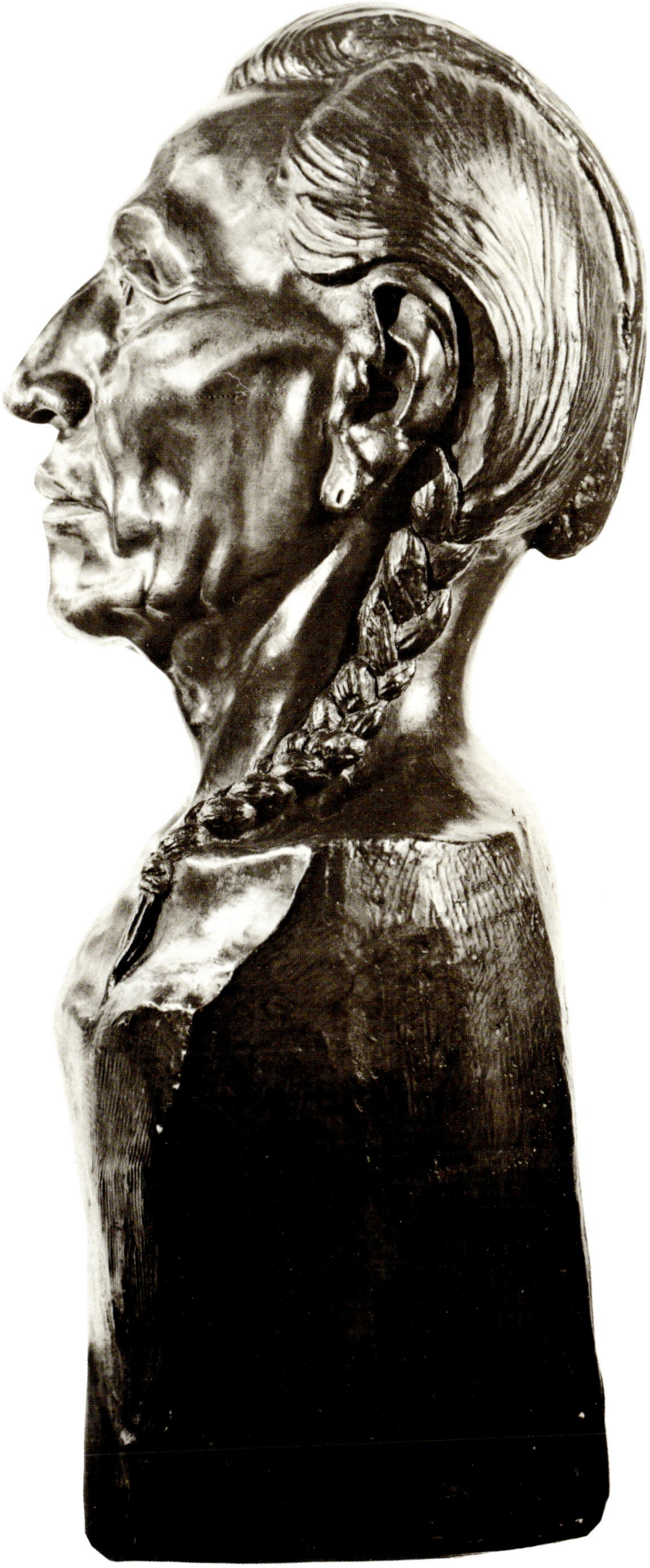

4 **SU-QUE-O-SIS-CON.** Big Spring. Blackfeet. Bronze Height 23 inches. Dated 1929. Chief Big Spring, a Full-Piegan, was known as a mighty hunter, and was feared by many because of his size. He stood 6′-5″ in his moccasins. Son of Wolf Bedding and Many Necklace, he married No Owl Woman, they had no children, she died. Then he married Small Face, all their children died, as did she. His third marriage was to Rabbit Old Woman, they had a girl, Medicinechild, and a boy, Dont't Talk. He was 64 in 1929 at the time he was sculpted. Big Spring's son, a full blooded Blackfoot, is now on the board of the Museum of Native American Cultures, a member of the National Republican Party Committee. He is also a commissioner of Glacier County, Montana, and he lives in East Glacier where he is the owner of a large cattle ranch.

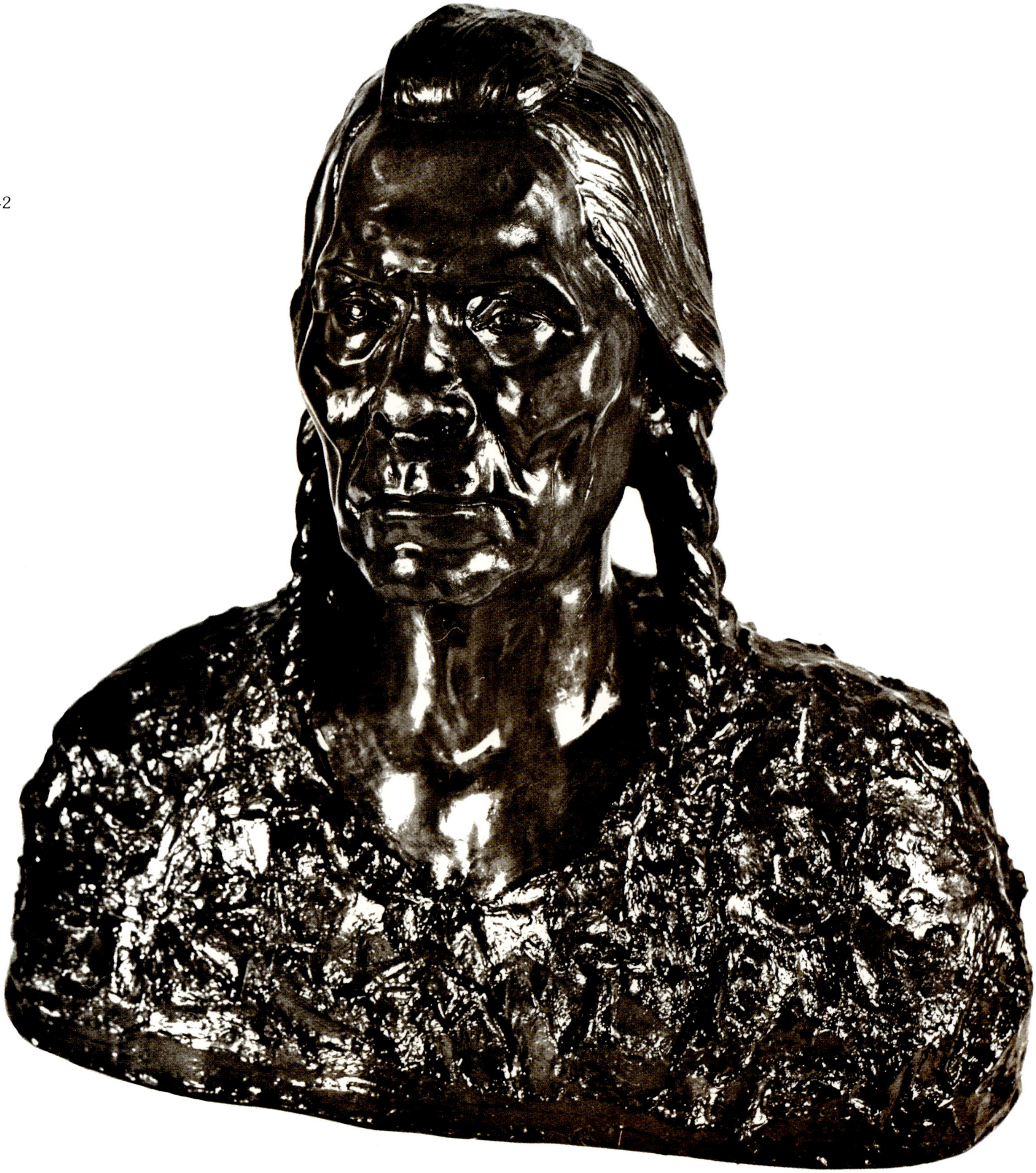

5 **E-TA-NA-MAKA.** Dog Gun. Blackfeet Brave. Bronze Height 24 inches. Dated 1929.

He was the son of Spotted Head and Wild Cat Face, and had a reputation for his accuracy with his saddle gun, and famous for his amusing story-telling ability. In 1929, at the time of this sculpture, he was 42 years old.

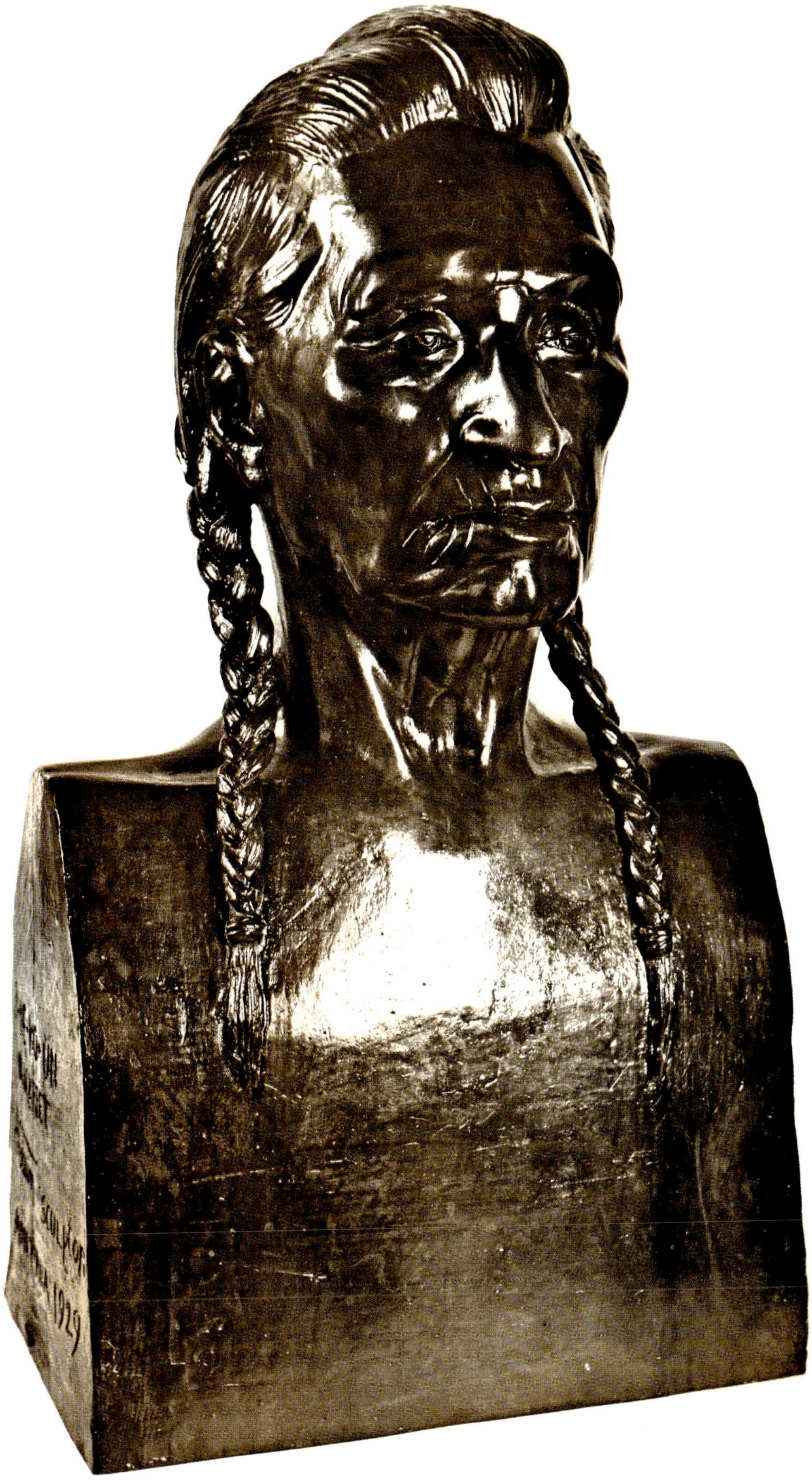

6 **KI-NE-KO-UN.** Jim Blood. Blackfeet Brave. Bronze Height 25½ inches. Dated 1929.
Jim Blood as a young man lived for 5 years with the Crow Indians. After leaving their camp he led numerous raids for horses against them.

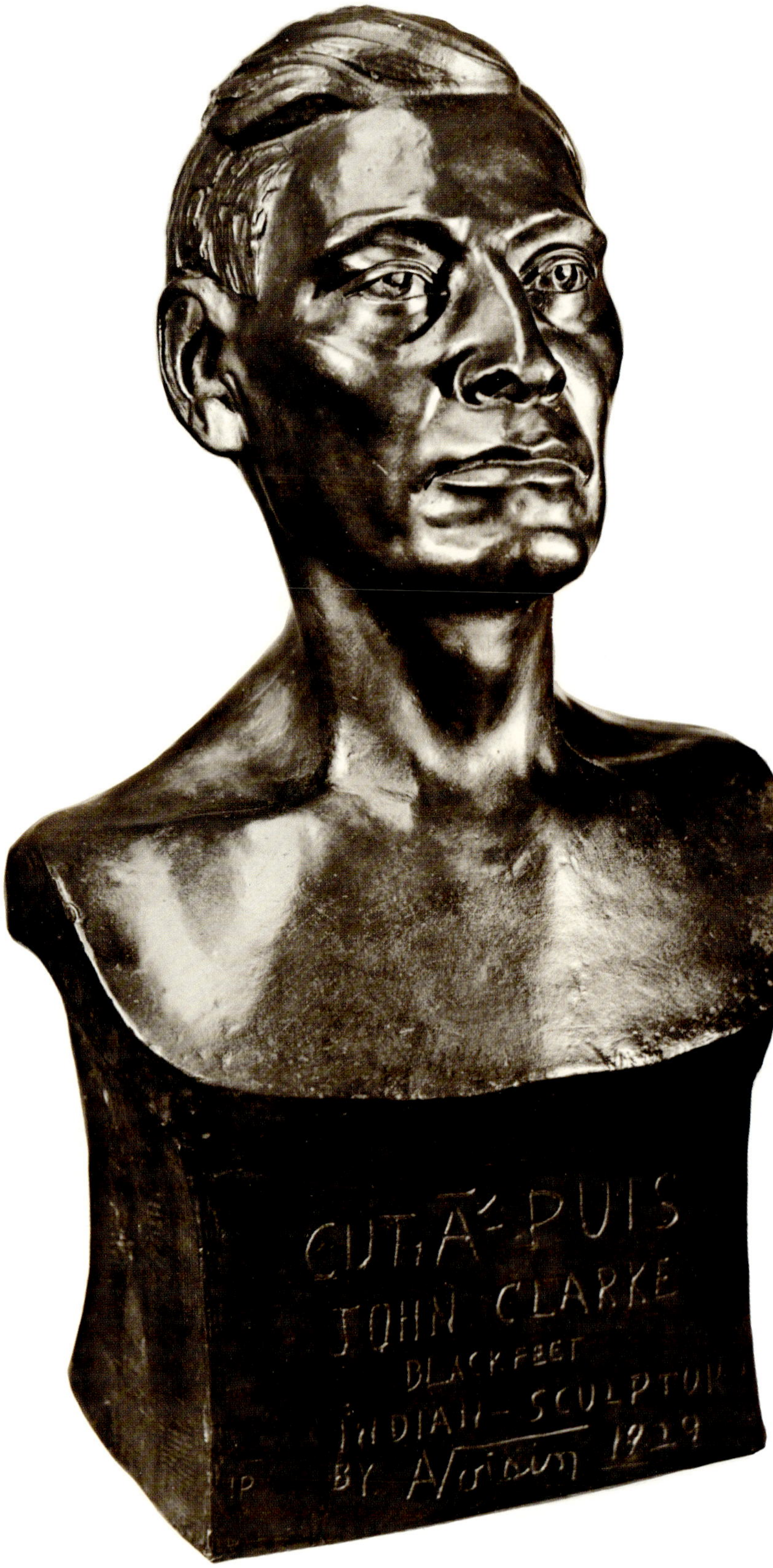

7 **CUT-A-PUIS.** John Clarke. Blackfeet. Bronze Height 25 inches. Dated 1929
A deaf-mute, and great at sign languages, which he developed from his knowledge of Indian inter-tribal languages and from writing notes. He was ¾ blood and the grandson of the white Malcolm E. Clarke and Cutting Off Head Woman (full Piegan). Malcolm Clarke was the rancher that Mountain Chief's band murdered in 1869, precipitating the 1870 massacre of friendly Indians at Marias Creek by U.S. Cavalry. During the raid on Malcolm's ranch (the Blackfeet say it was a feud culmination, not an indiscriminate killing), John's father, Horace J. Malcolm, was shot in the face and left for dead. Adrien's wife, Francis, helped John Clarke by interpreting the lessons on carving that Adrien taught. With unusual ability, CUT-A-PUIS was quick to grasp this form of art, and in a short period of time became an accomplished wood carver. He went to Minneapolis to further study carving and sculpturing. Later he set up his studio in East Glacier, Montana, where he would sell his carved figures to the tourists that were brought in by the railroads. His reputation grew, and his art is sought after by collectors. At the time Voisin made his sculpture in 1929, John was 49.

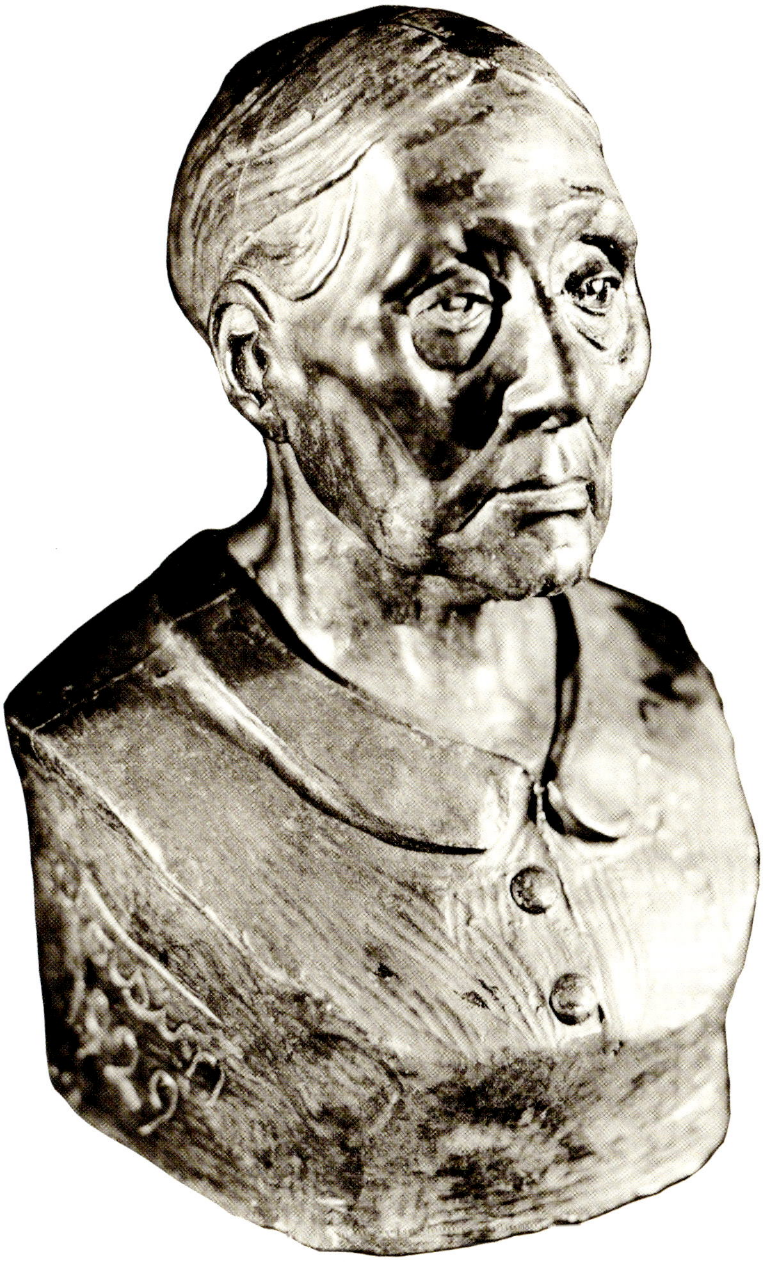

8 **FIRST KILL.** Full Blood Piegan. Bronze Height 8 inches. Dated 1929.
The daughter of Mink and Curlew, she was CUT-A-PUIŚ mother. At the time her sculpture was made by Voisin in 1929 she was 79 years old.

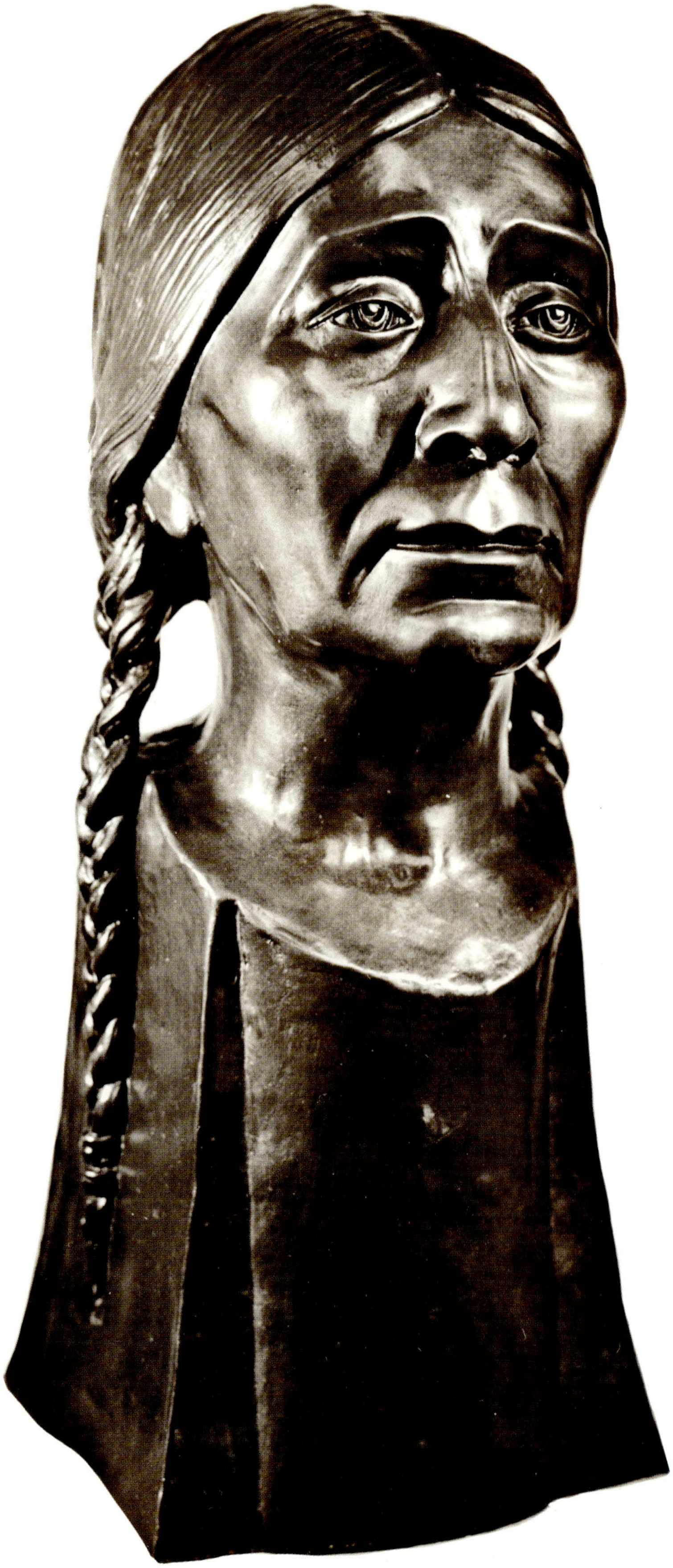

9 **MA-MEIN-E-MA.** Scalping Woman. Blackfeet Squaw. Bronze Height 21½ inches. Dated 1929.

So called the woman with "the flathead," because her mother was a Flathead Indian. Scalping Woman was the wife of Nightshoots.

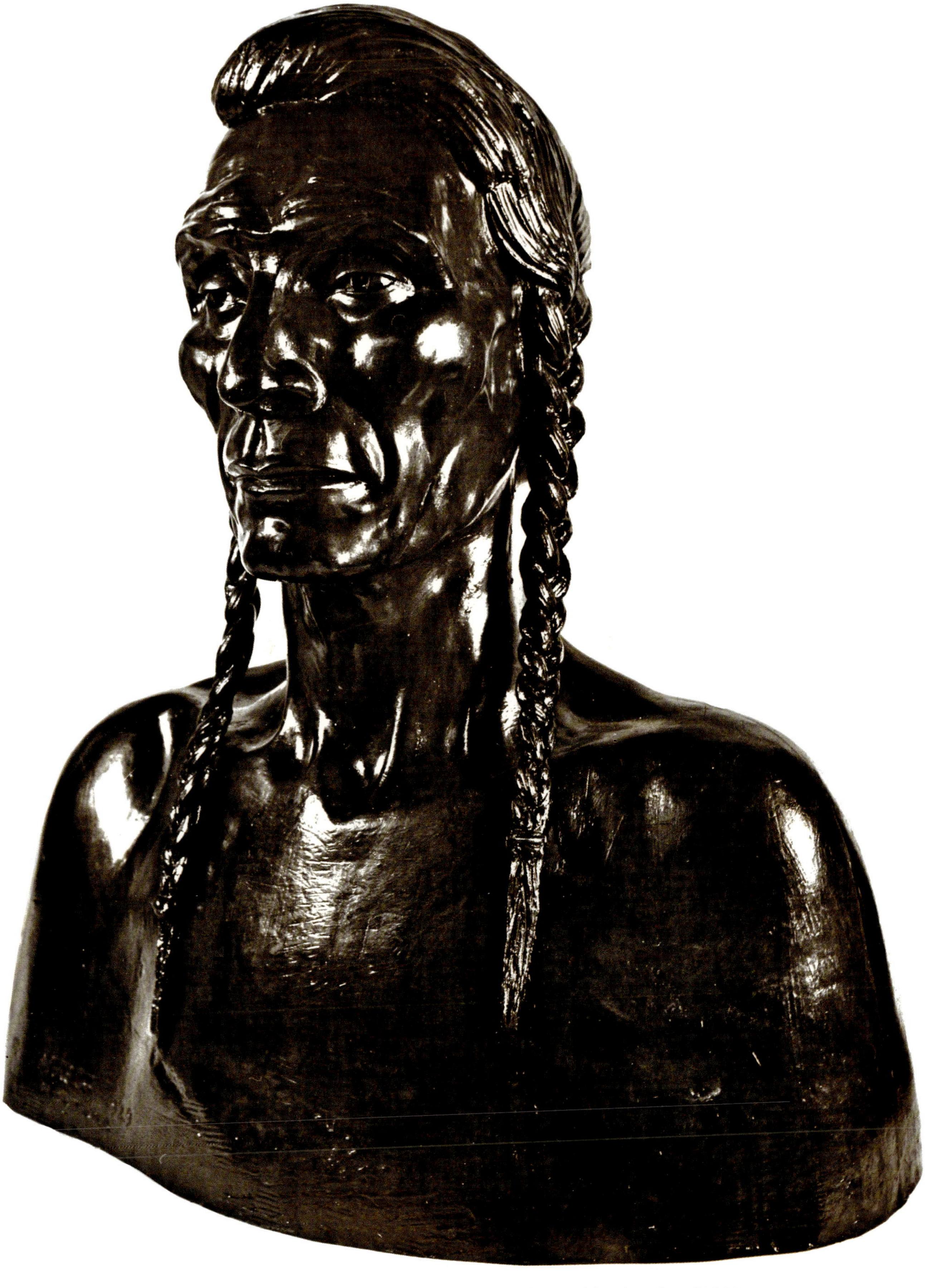

10 **IN-A-SI-KIN.** Buffalo Hide. Full Piegan Bronze Height 24 inches. Dated 1929. The son of Blackfeet Woman and Hansome Man, he married Spotted Woman when she was 15 years old.

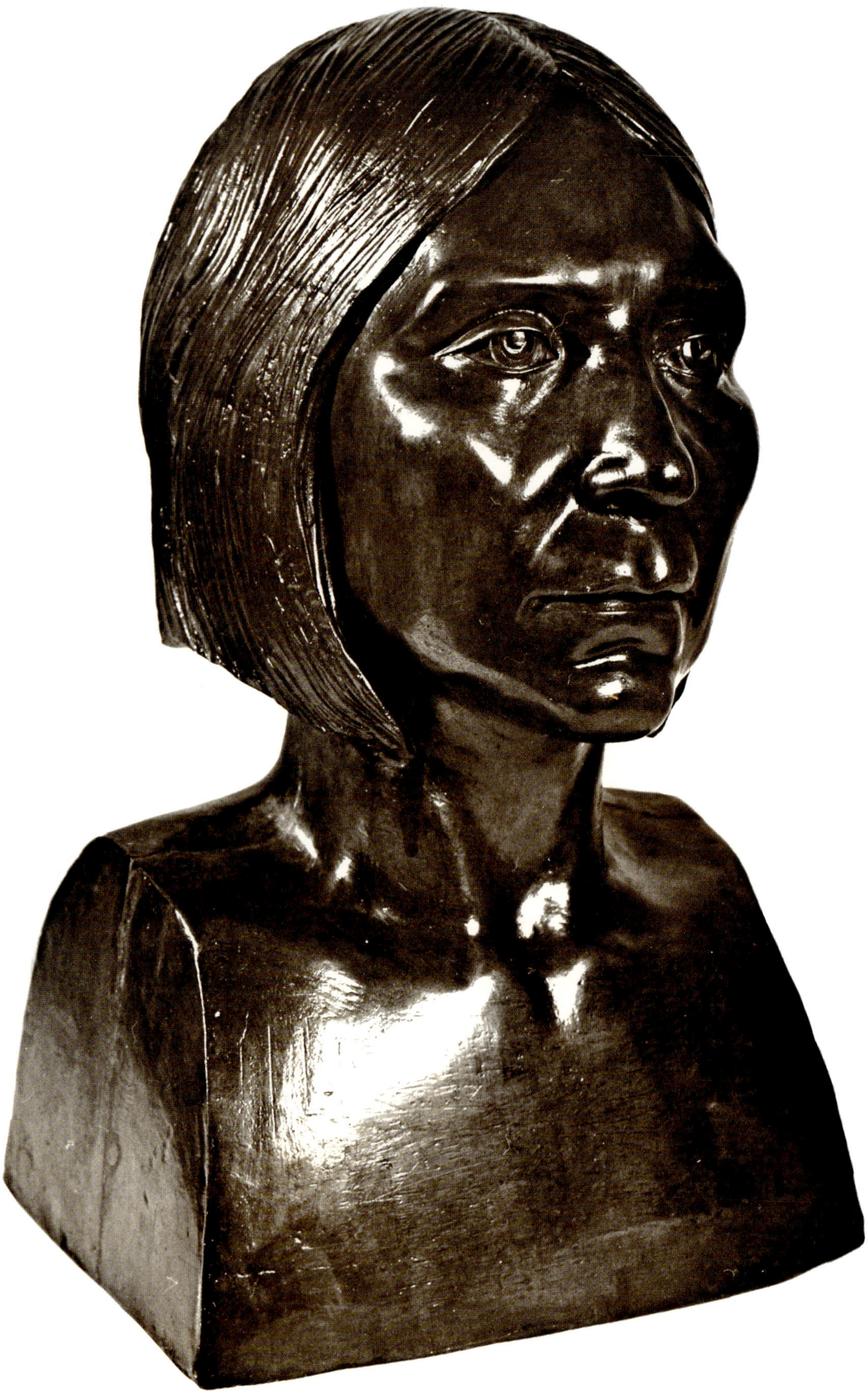

11 **BERTHA.** Blackfeet Squaw. Height of Bronze 17 inches. Dated 1929. Daughter of Many Tail Feathers, and Long Time Bear Woman, was married at the age of 10 to Louis Plume. When her sculpture was made in 1929 she was 42.

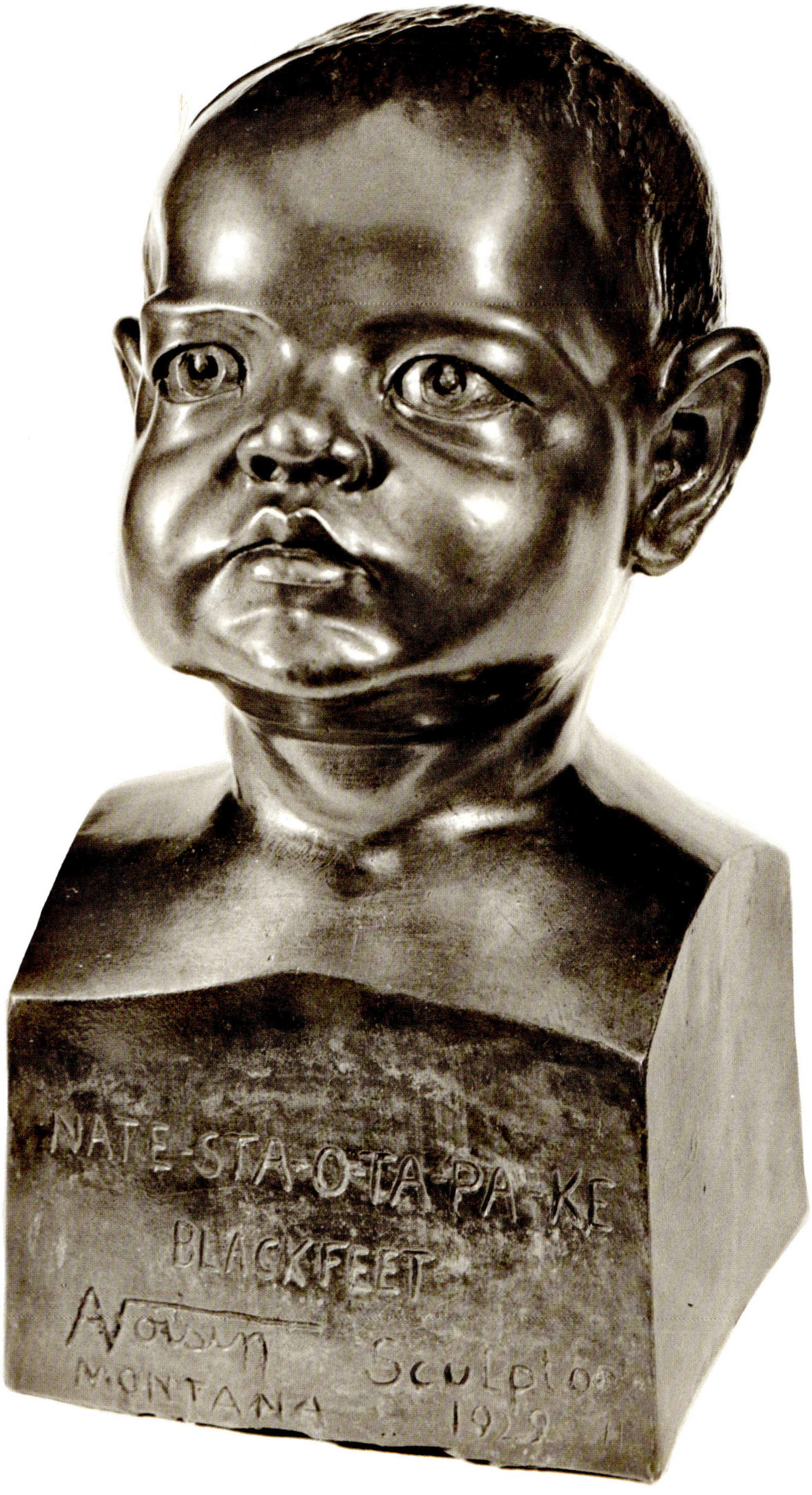

12 **NATE-STA-O-TA-PA-KE.** Blackfeet Papoose. Bronze Height 11¼ inches. Dated 1929.

13. Indian Nativity. Although Voisin professed no specific faith, he had a profound feeling and respect for religion, whether it was Indian worship or White Man's beliefs. He recognized that everyone carried within himself a spiritual commitment to a higher order. The Indians respected him, for he always showed them as a peaceful people, and in this section, carved on slate, he portrays the "true inner feeling" of the American Indian that he came to know and understand.

The Indian portraits in the Indian Nativity Bas-Relief are: l to r Cecile Last Star, Blackfeet; Nina-Stuko, Blackfeet; Talla-We-Ton-Me, Nez Perce; Nate-Sta-O-Ta-Pake, Blackfeet Papoose; Su-Que-O-Sis-Con, Blackfeet, and In-A-Si-Kin, Blackfeet.

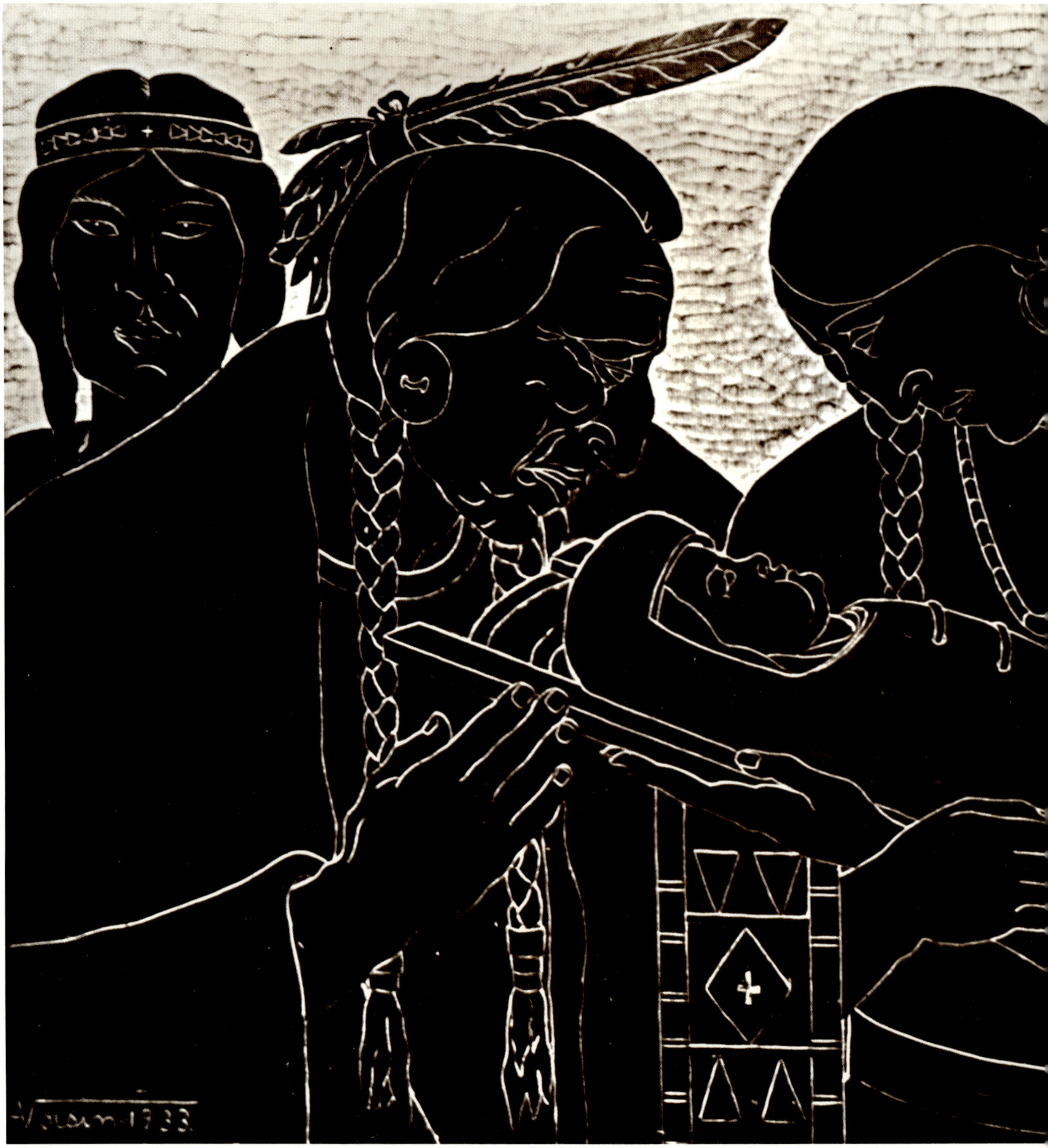

NEW MEDIUM

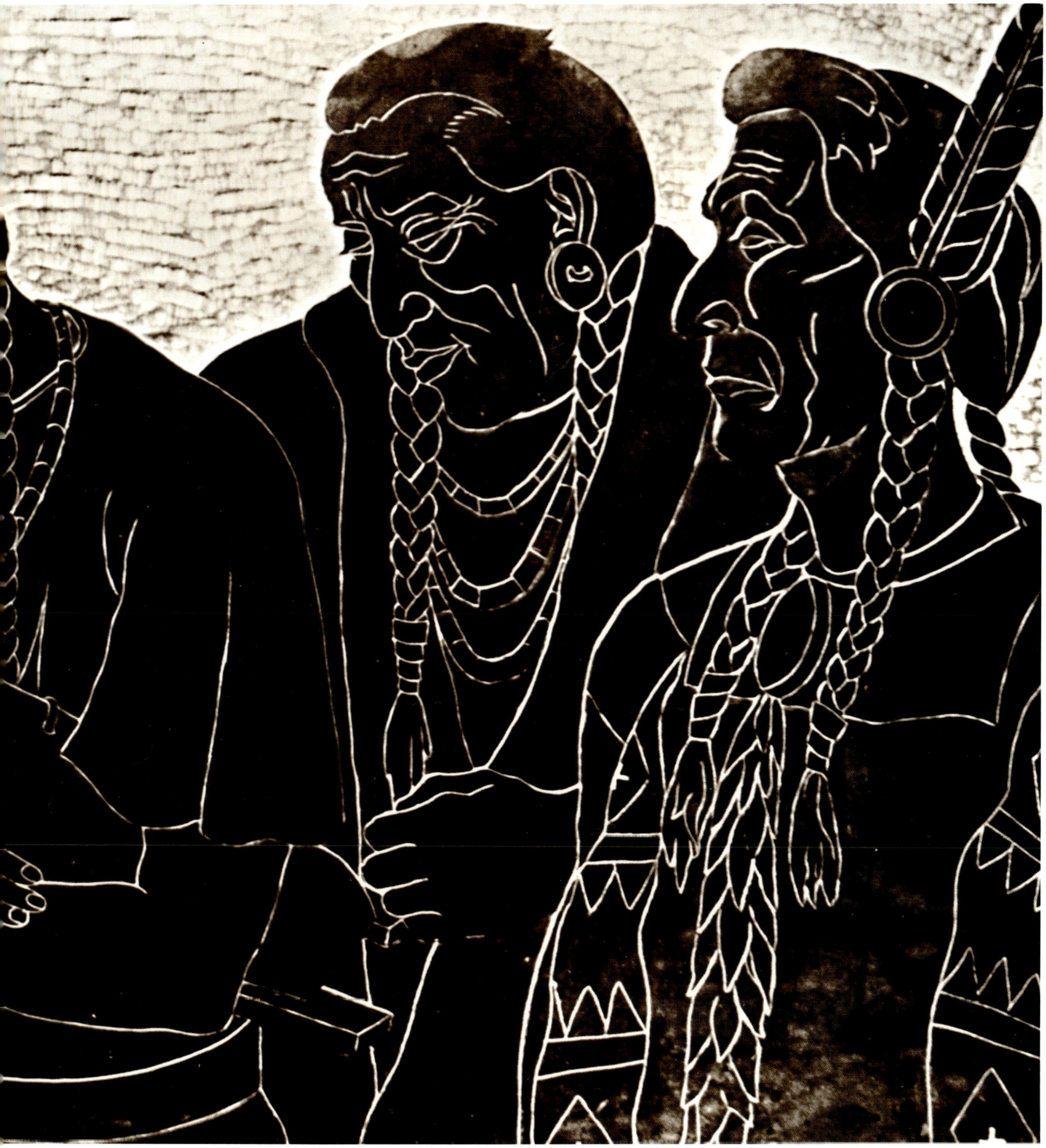

SLATE

Adrien Alexandre Voisin not only lived among the Blackfeet Indians, but also participated in their customs and ceremonies as an honorary tribesman.

Manifested in these bas reliefs is an intense concentration on personal dignity and focus on racial pride. As in the essence of all ceremony and tribal fortitude, Voisin presents a sculpture to immediate response and perfection.

Upon his return to America in the early 1930's, Voisin ventured into a new technique, practically unknown to the art world, sculpture on slate. After many unsuccessful attempts, he formed a special backing embedded with steel rod that stabilized the strength of the slate and helped keep this brittle material from cracking. After hundreds of hours of hand carving and using a chisel and mallet and with much work on hand rubbing the finished surface, he successfully executed the three shown.

These slates are beautiful, but so fragile. The bronze castings made from the slates are likewise magnificent, but will endure the ravages of time and travel.

Since the Harmsens acquired these pieces, they have had the molds made and produced by the lost-wax process.

The large scene of the Indian nativity was produced in 19 separate parts and then welded, chased and painted.

The "Les Voyageurs" in 6 pieces and "Two Gun White Calf" in 4 pieces.

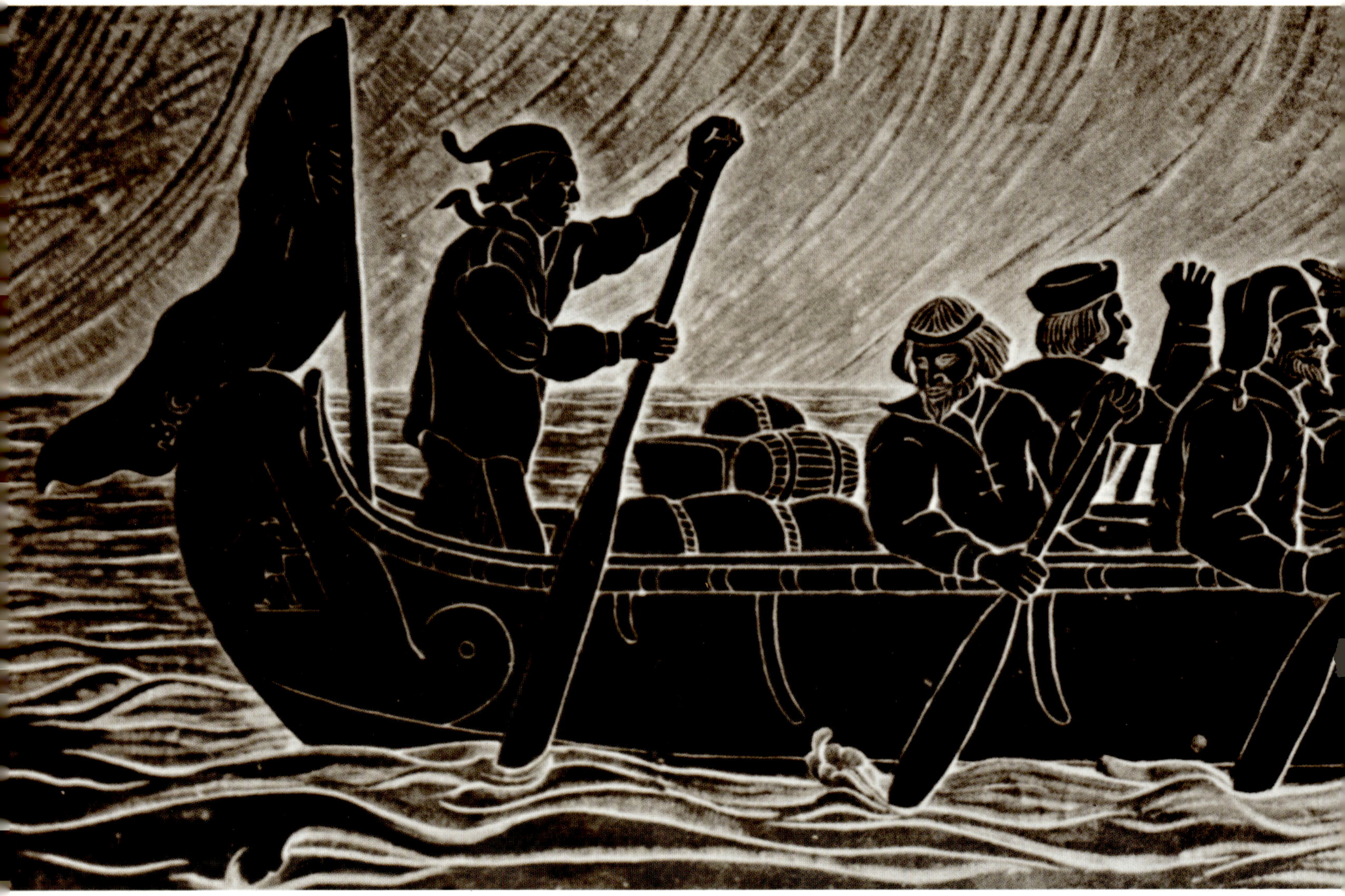

15. Dr. John McLoughlin "LES VOYAGEURS"—This bas-relief slate depicts the arrival from Canada of a canoe paddled by a crew of eight members of the Hudson Bay Company at a port along the Columbia River.

The Man, Dr. John McLoughlin, was born in Parish La Rivare du Loup, Canada, October 19, 1784. Known eventually as "the father of Oregon," he died September 3, 1857, in Oregon City. His life was an essential part of the early, heroic days of early Oregon.

In 1821, he was put in charge of Fort William, Canada, the chief depot and factory of the Northwest Company, when the company merged with the Hudson Bay Company. He left Canada for Oregon in 1824 by boat on the Columbia River, then went part way overland to Fort George (Astoria,) and after careful surveys, went up the Columbia to found Fort Vancouver. Here he was in charge of operations for the Hudson Bay Company.

In 1848, Dr. McLoughlin resigned from the Hudson Bay Company, when he had been advised, in a letter from Governor-in-Chief Simpson of that company that he should no longer assist the needy. His reply was, "Gentlemen, if such is your order, I will serve you no longer."

On May 30, 1849, McLoughlin took the oath to become a citizen of the United States as required by Naturalization Law. October 6, 1905, was set aside as McLoughlin Day by the Lewis and Clark Exposition, Portland, and later, Oregon boys and girls, through popular subscription, erected the McLoughlin Shrine at Oregon City, Oregon.

Adrien Voisin executed a large bust of McLoughlin in bronze, which now stands near the latter's old home in Oregon City. The artist was so impressed with his deeds and humility that he put in slate a permanent memory of this remarkable man, who had done so much to help the early pioneers. It was cast in bronze in 1978 to preserve it and to commemorate the work of Sculptor Voisin, and the many sacrifices of Dr. John McLoughlin.

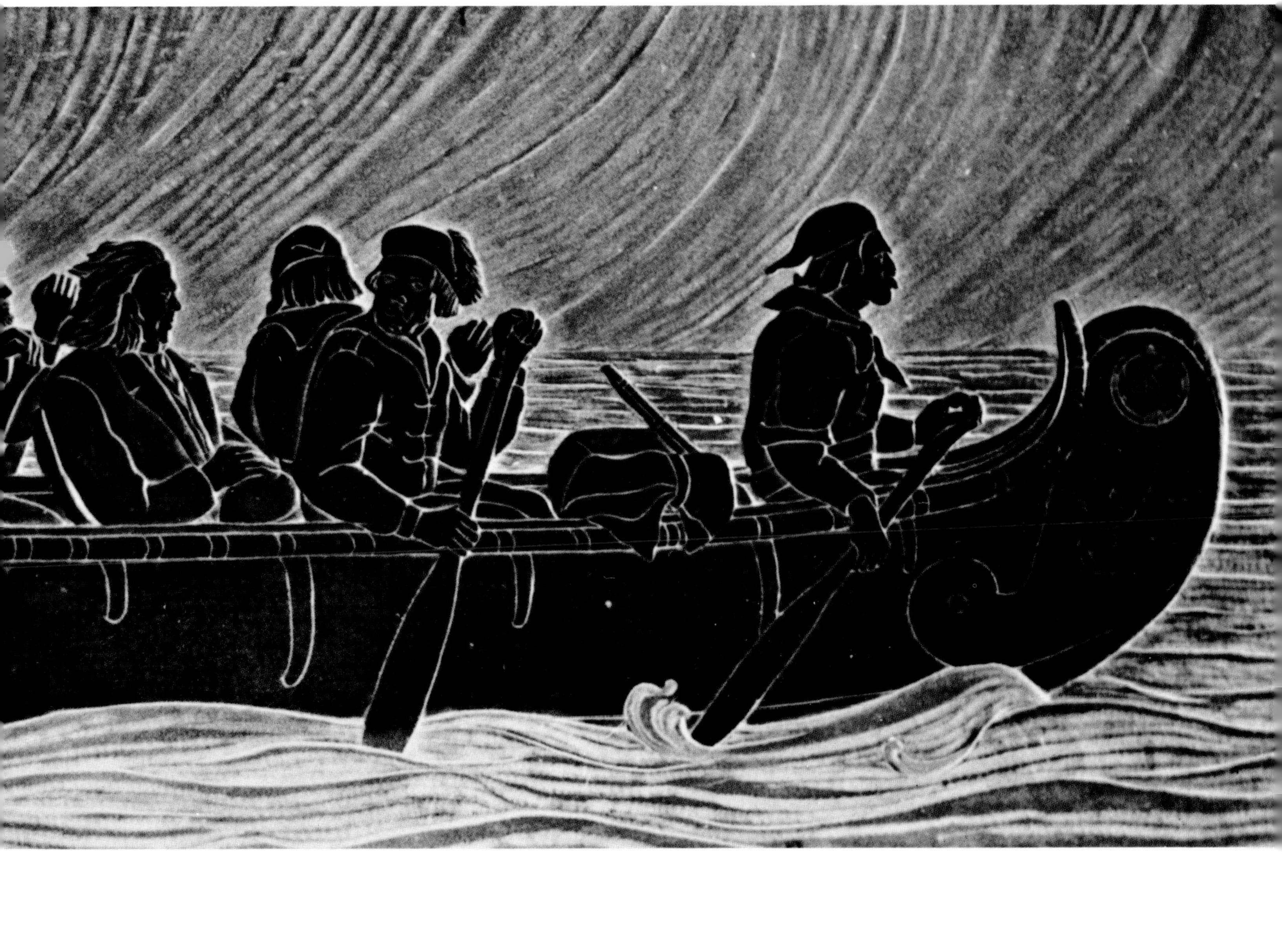

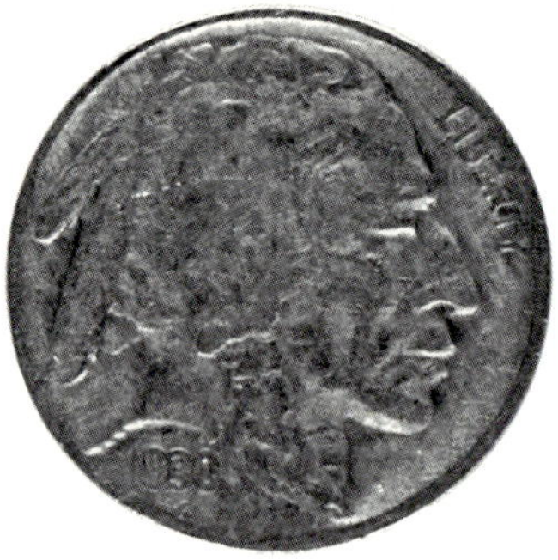

James Earl Fraser designed the five-cent "Buffalo Nickel" that was issued in 1913. He used three different Indians to obtain the portrait. Two of them were, "Two Guns White Calf," and Sioux Chief "Iron Tail."

The bison on the reverse side was modeled after Black Diamond in the New York Zoo, that was slaughtered in 1915 and mounted.

14. Two Guns White Calf. One of Adrien Voisin's most admired friends. Voisin and his wife, Mary, were made members of his tribe by Chief Two Guns, and given the Indian name of AH-GAW-CHE-NENA. Voisin made many portraits of this famous Indian. The friendship was mutual. Chief Two Guns not only made him a friend and a tribe member, but gave Adrien his peace pipe and quilled pipe bag.

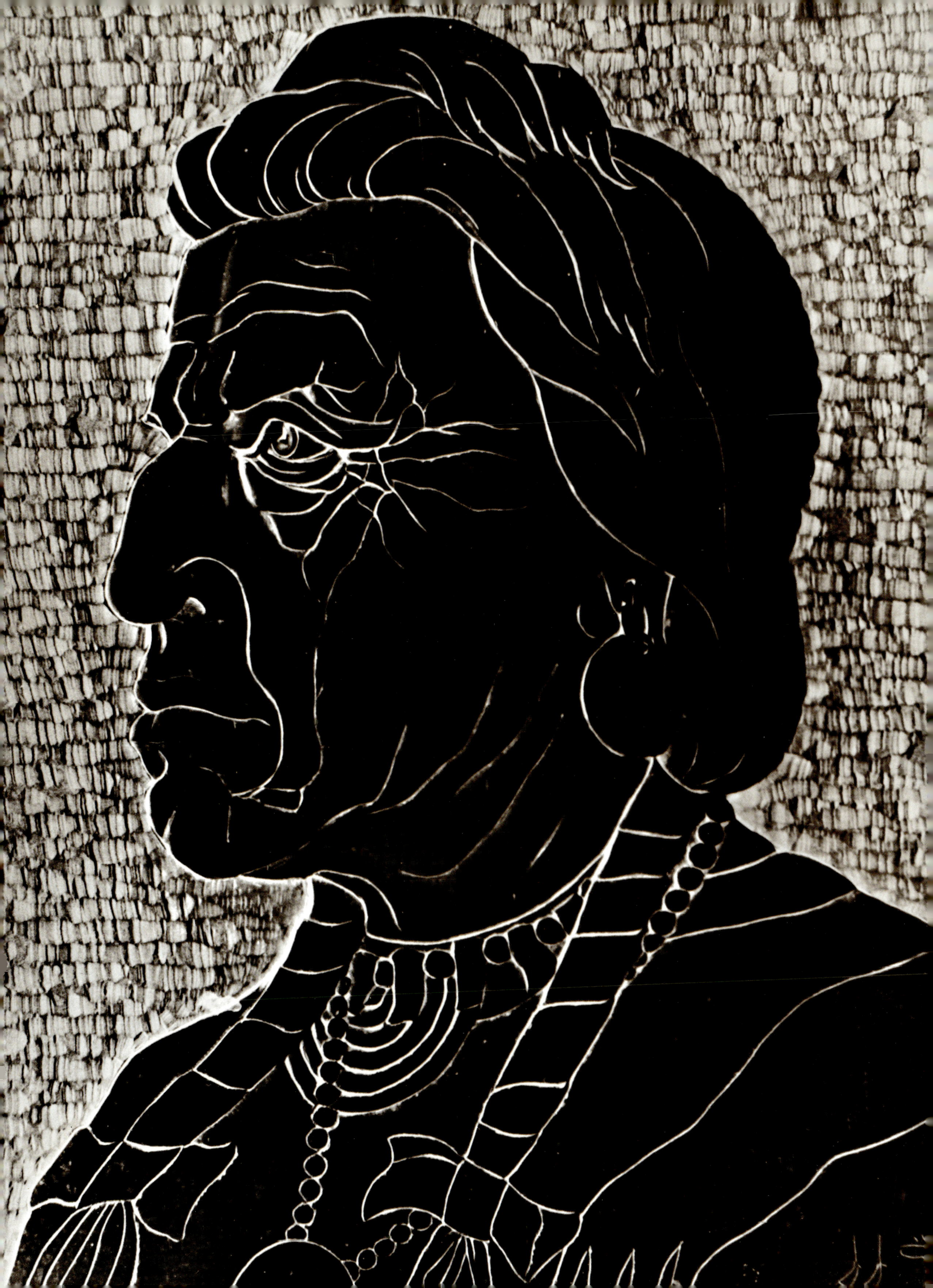

THE BLAC

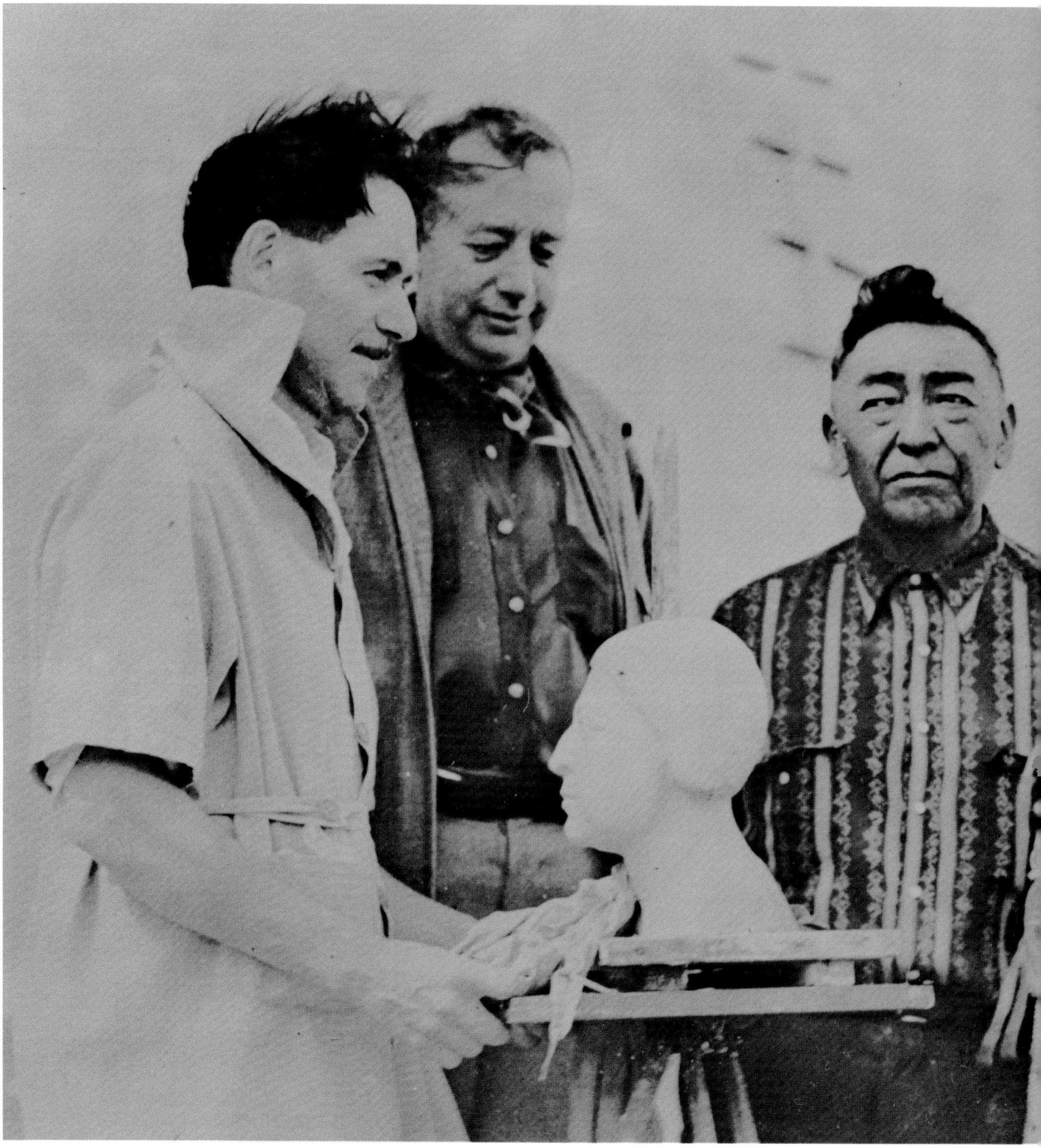

Voisin's Indian Gallery Captured in Bronze

KFEET INDIANS

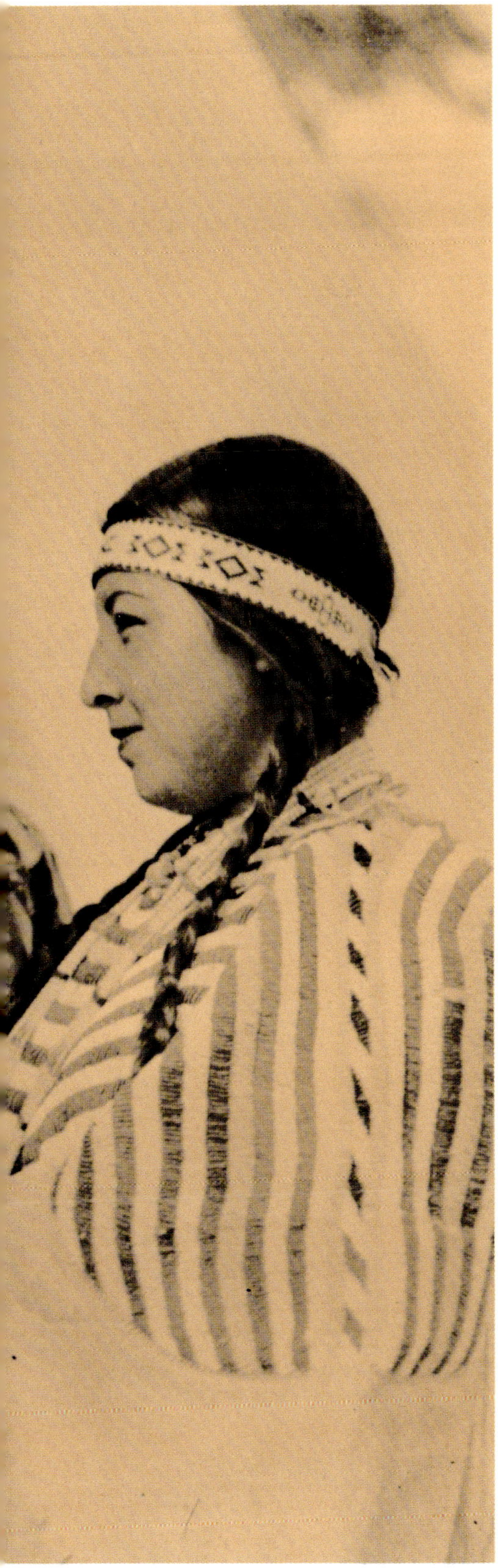

The early Blackfeet Indians of Alberta, B.C., and Montana, U.S.A., are synonymous in their cultures. The Piegan (Pikuni,) the Bloods (Kainow,) and the Blackfeet (Siksika) are closely allied Algonquin tribes collectively designated as "Blackfeet." Before they migrated westward to the plains of Montana and Alberta, they were forest Indians. These tribes struggled against the invasion of the white man with a determination that has not been equalled in the annals of Indian warfare, and it is only within recent years that they have accepted the inevitable."

Living in a manner as primitive as that of the early cavemen of Europe, they ate the same food; killed similar varieties of game with identical weapons; and built fires no differently from their neighbors. The buffalo furnished their principal food supply, supplemented by elk, deer, and antelope, with the addition of roots and berries. They planted small amounts of tobacco for ceremonial and religious purposes. But they did not grow vegetables or grain. They were, and still are, meat eaters.

The skin, bones, and horns of the animals supplied material for their dwellings, implements, and clothing. Pottery and metal utensils were not used.

Cooking was a remarkable process. Digging shallow holes in the ground, they lined them with pieces of green hide, fur side down, and then dropped heated stones into them. When the water began to boil, the meat was added. Salt was not used extensively. Meat was roasted embedded in hot stones, also.

It is an established fact that distilled, intoxicating liquors were never brewed.

Physically, the Piegan are of medium stature, of a firm, compact build. The squaws look quite masculine because of the heavy work required of them in childhood.

Polygamy was the rule, but the first wife was considered the "real" wife. The other wives were considered as servants to the first, and in many instances, they were her younger sisters. If an unfaithful wife were not killed, she had her nose or ears cut off as an object lesson to others.

Unfortunately, only a few fullbloods exist today. Trichoma is fatal to their sight, and many of the elders are nearly blind from this disease. Intermarriage with the whites is causing the rapid disappearance of this noble tribe of plains Indians.

L to R. Adrien Voisin, sculptor, a writer, unidentified, a Nez Perce Indian father and his daughter, looking over a sculpture that Adrien made of the daughter.

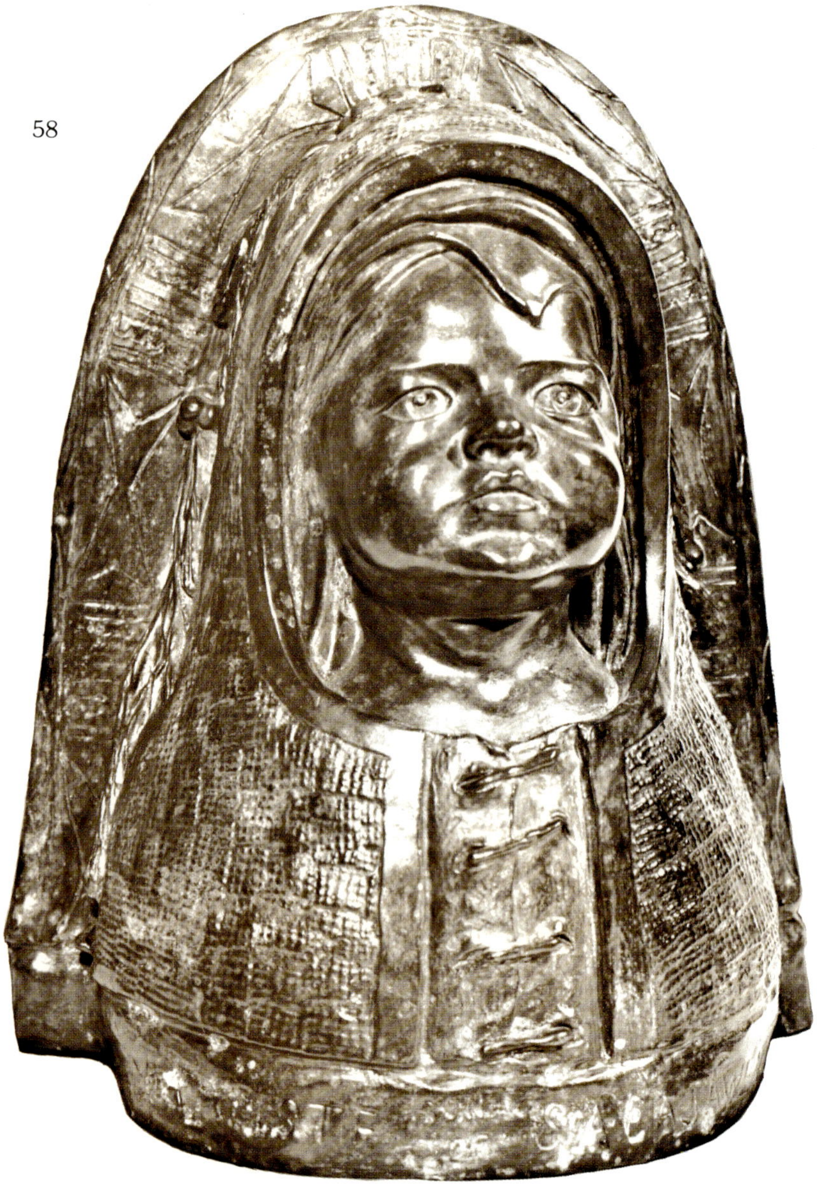

16 NATE-STA-O-TA-PA-KE,
Papoose Baptiste, Blackfeet. Son of Sacajawea. Bronze Height 15 inches. Dated 1927.

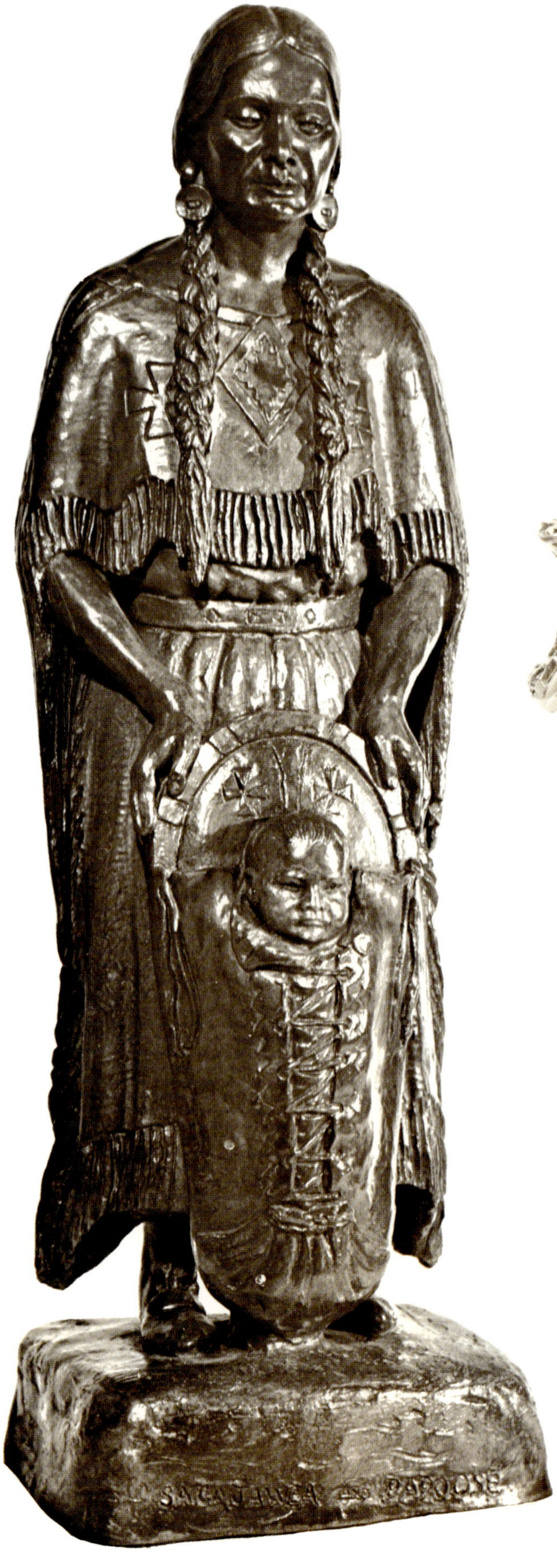

17 SACAJAWEA,
Melissa Parr. Squaw with Papoose, NATE-STA-O-TA-PA-KE. Bronze Height 20 inches. Dated 1936.

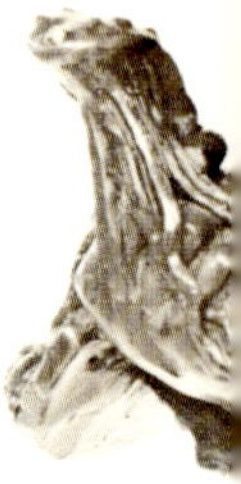

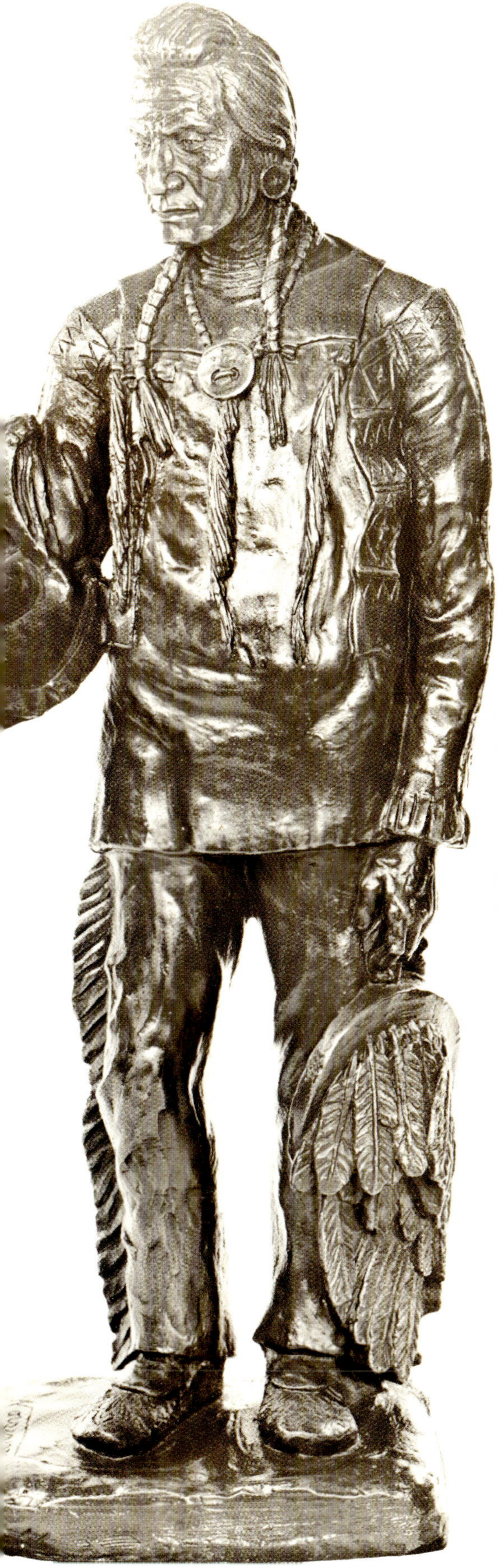

The Indians represented in this exposition belong to the Blackfeet tribe of North America. The name "Blackfeet" was given them by the pioneers, coming from the fact that the soil of the region they inhabited was composed of dark-colored clay, also very oily, which adhered to their moccasins. Their true name is "Picuni," and, although not numerous today, they were at one time a large, powerful tribe.

Their names, which seem strange to us when they are translated into the white man's tongue, are extremely harmonious and logical in the Indian language.

Most of the names were borrowed from visions in their sacred dreams. The animals and objects which appear in these dreams become for the Indian both a remedy and a talisman and are supposed to protect him against evil and against his enemies.

Sometimes, after cutting out V-shaped pieces from their ears, these Indians offered them in sacrifice during sacred ceremonies to the sun, hoping that such gifts would bring good luck.

Adrien Voisin found that hamburger meat was the greatest reward for the Indians who would consent to pose for him. Their hunting capabilities had greatly decreased because of Government regulations and the white man's plundering of their favorite game, the buffalo. In addition to meat, his gifts were cigarettes, beads, and money.

18 TWO GUNS WHITE CALF,
Blackfeet Brave, Montana. Bronze Height 20¾ inches. Dated 1933.

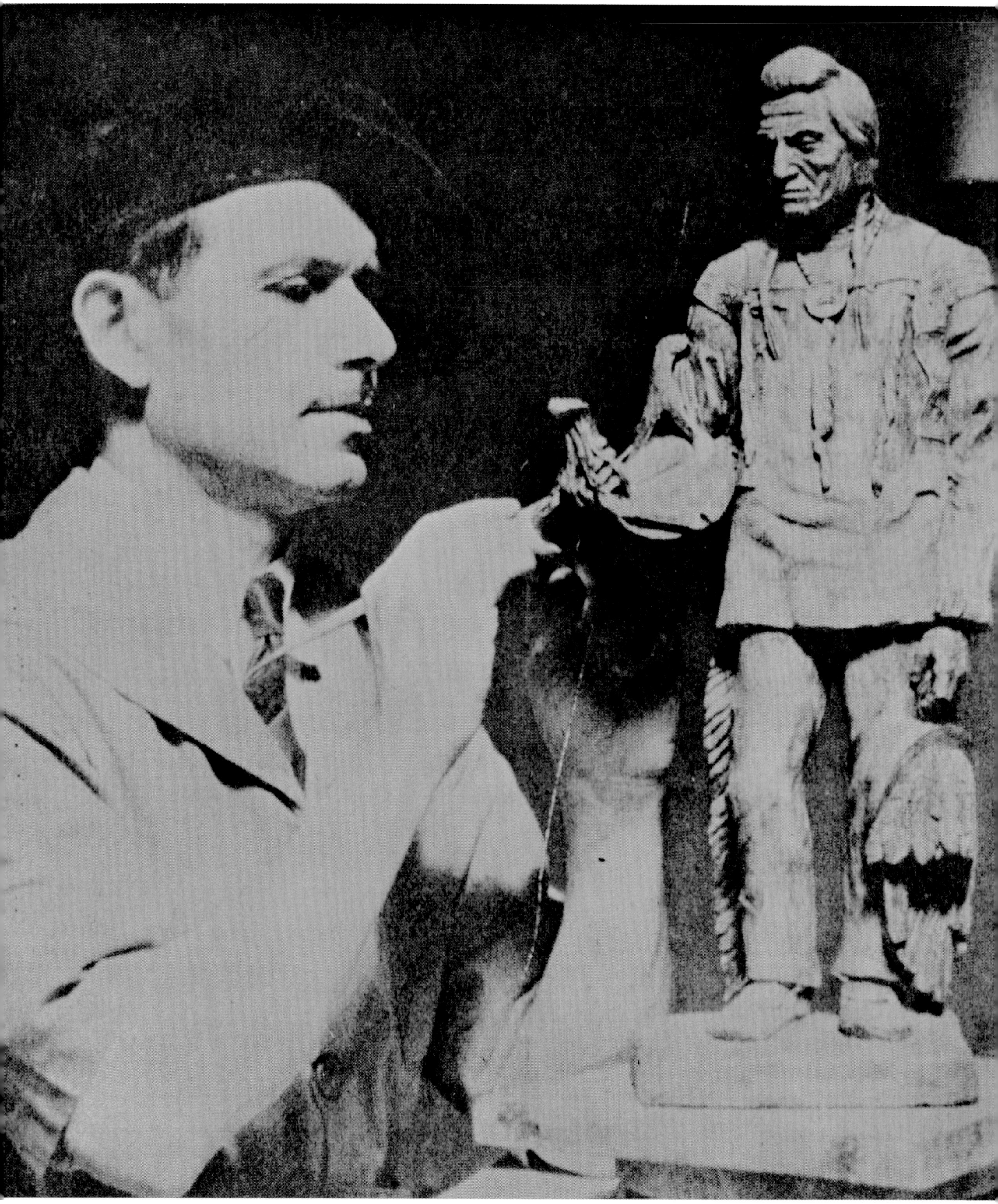

Adrien Voisin modeling Two Guns White Calf, Fort Browning, Montana, 1929.

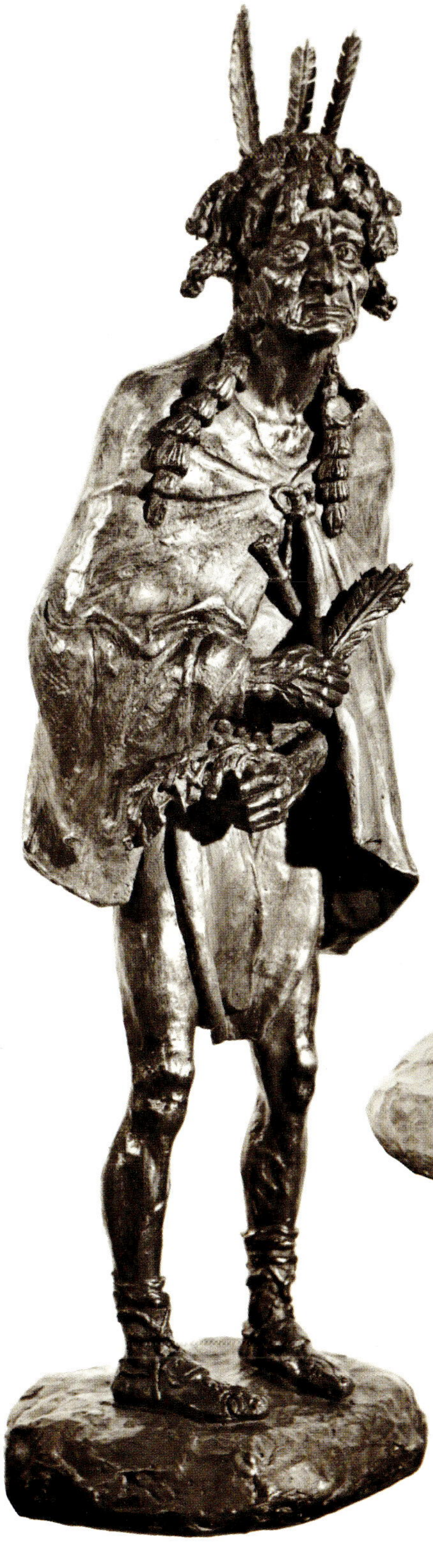

19 MEDICINE MAN,
Blackfeet. Bronze Height 21 inches. Dated 1933.

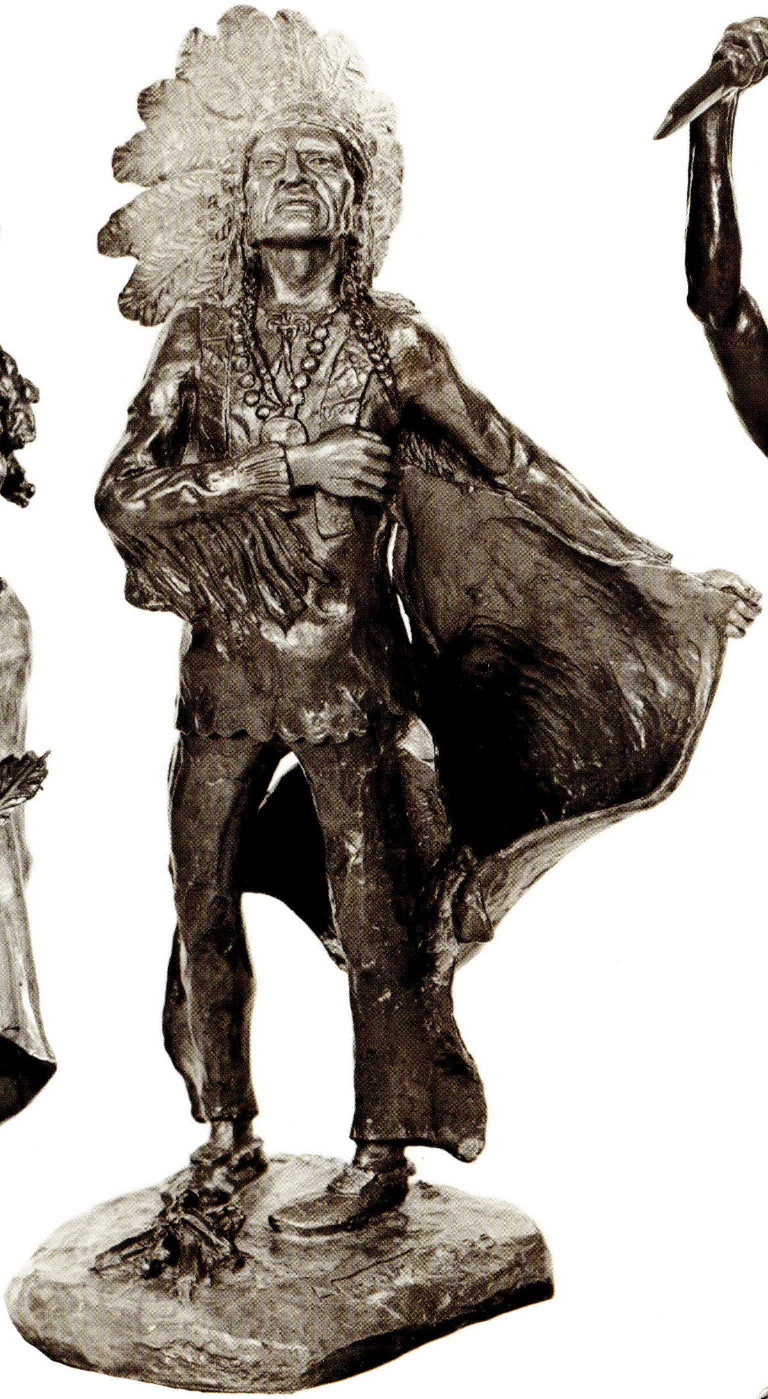

20 SMOKE SIGNALS,
Blackfeet. Bronze Height 12¼ inches. Dated 1935.

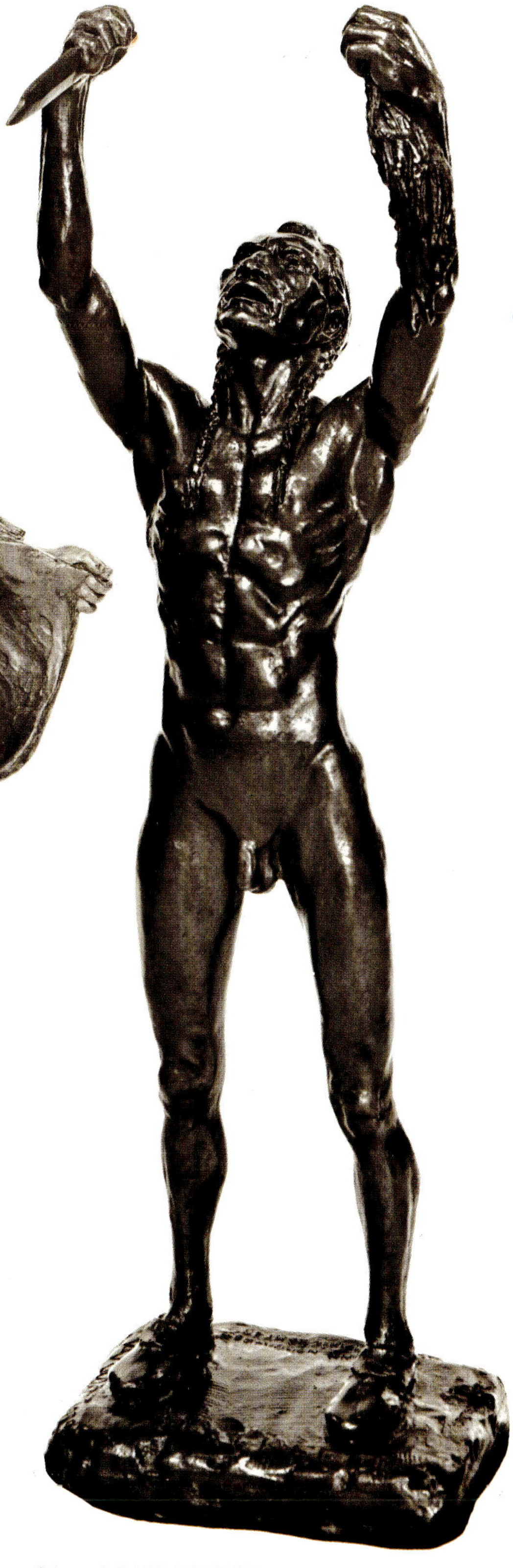

21 SCALP PRIEST,
Blackfeet. Bronze Height 15½ inches. Dated 1929.

62

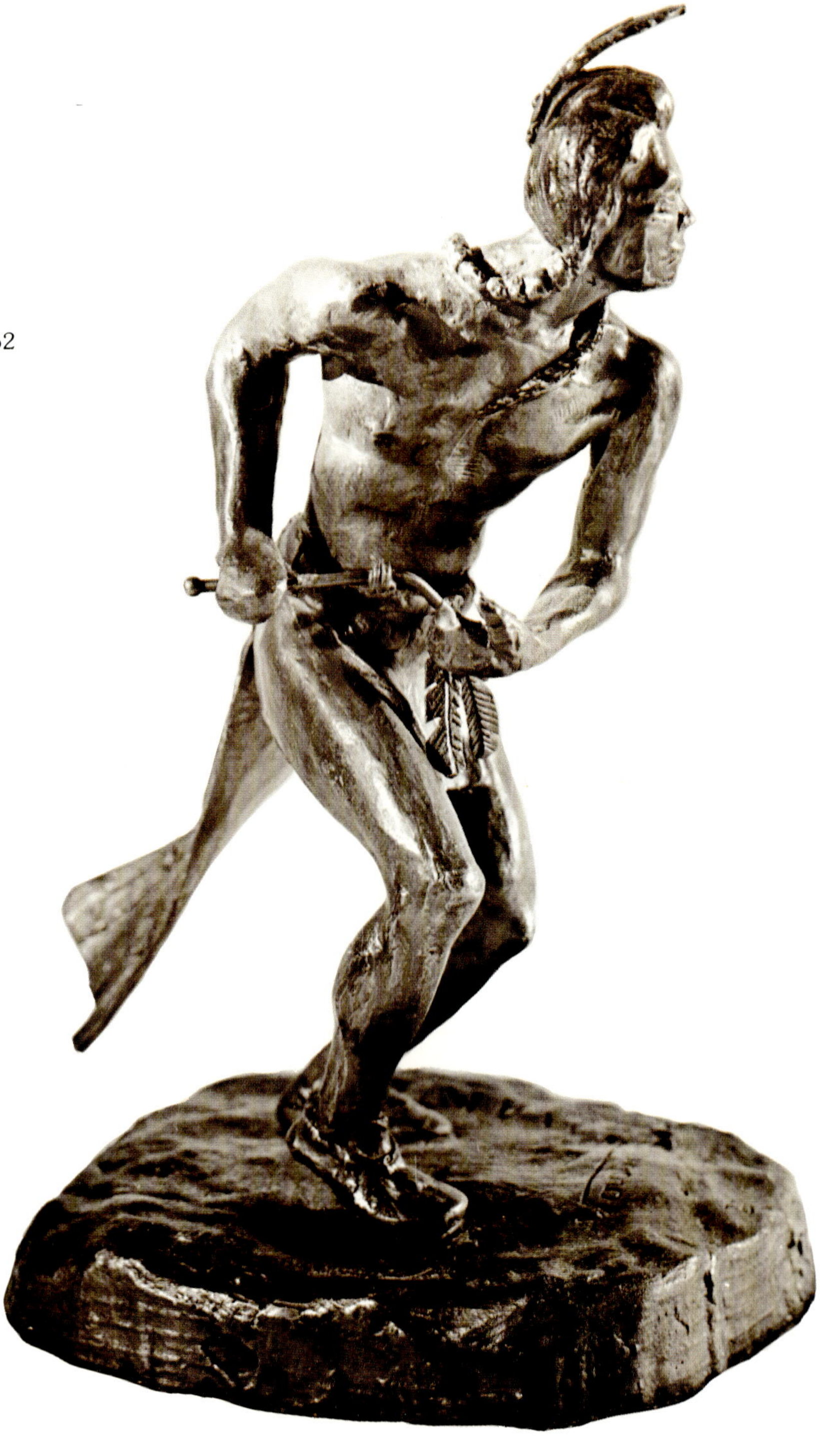

22 **YAKIMA DANCER.**
Bronze Height 9 inches. Dated 1936.

23 **CHIEF WITH PIPE,**
Blackfeet. Bronze Height 14 inches. Dated 1929.

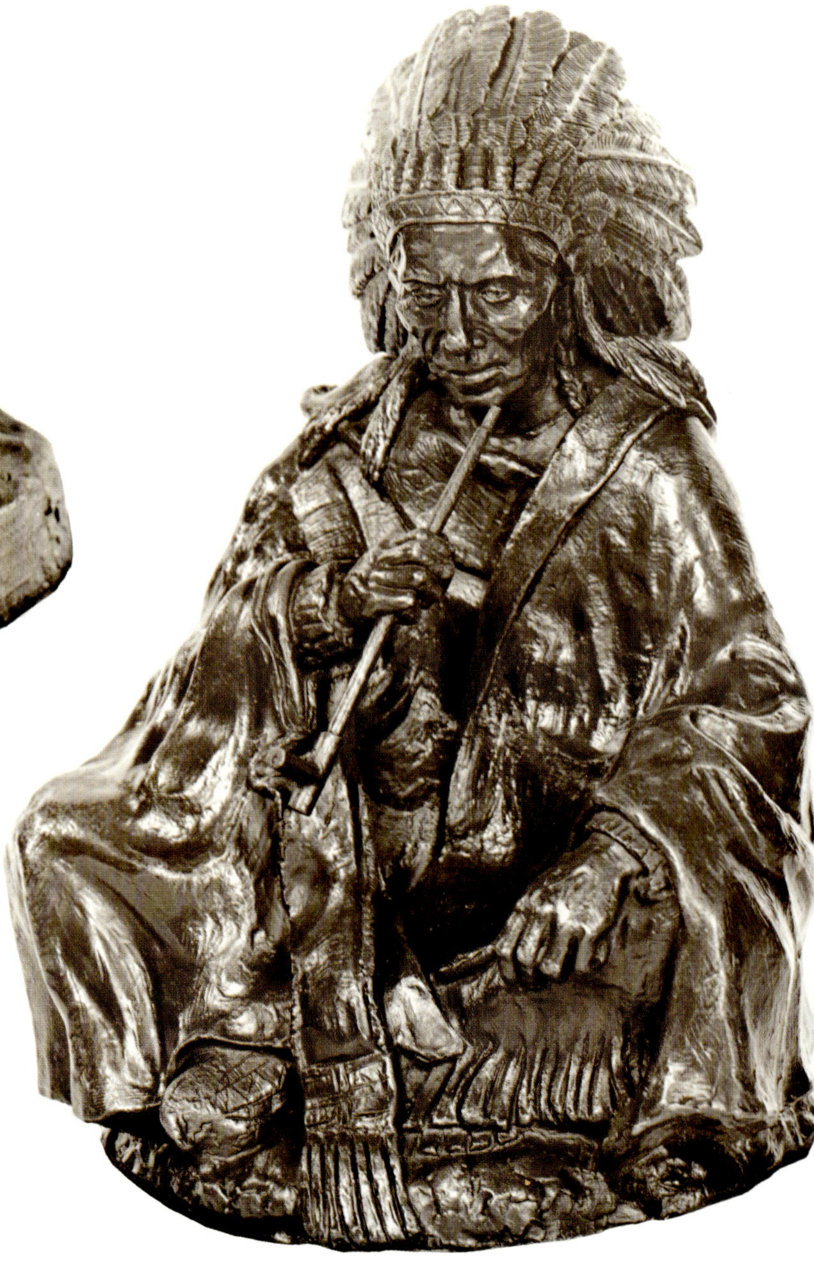

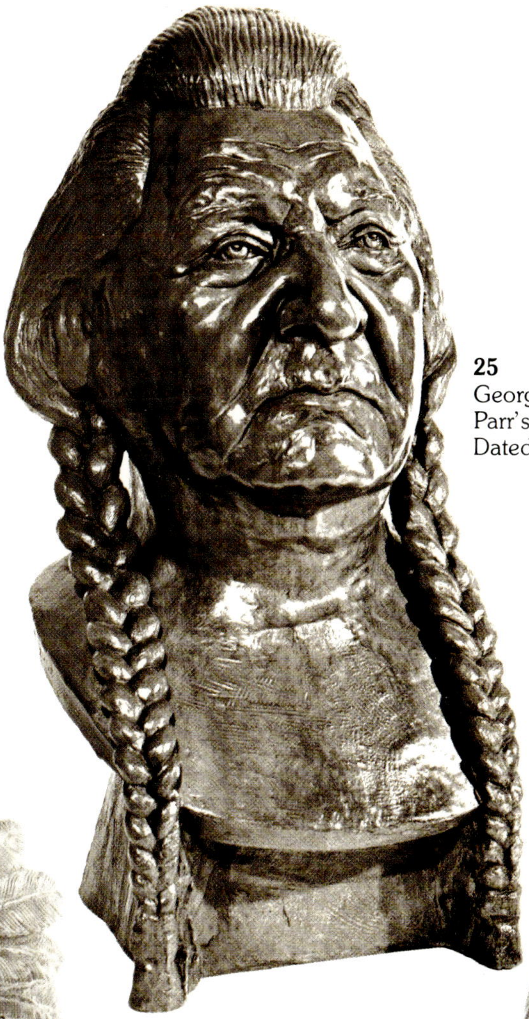

25 PE-TOM-YOM-NON-IL-PILP,
George Red Hawk. Cayuse Chief. Melissa Parr's Uncle. Bronze Height 19¾ inches. Dated 1936.

24 BUFFALO MEMORIES,
Blackfeet. Bronze Height 14 inches. Dated 1929.

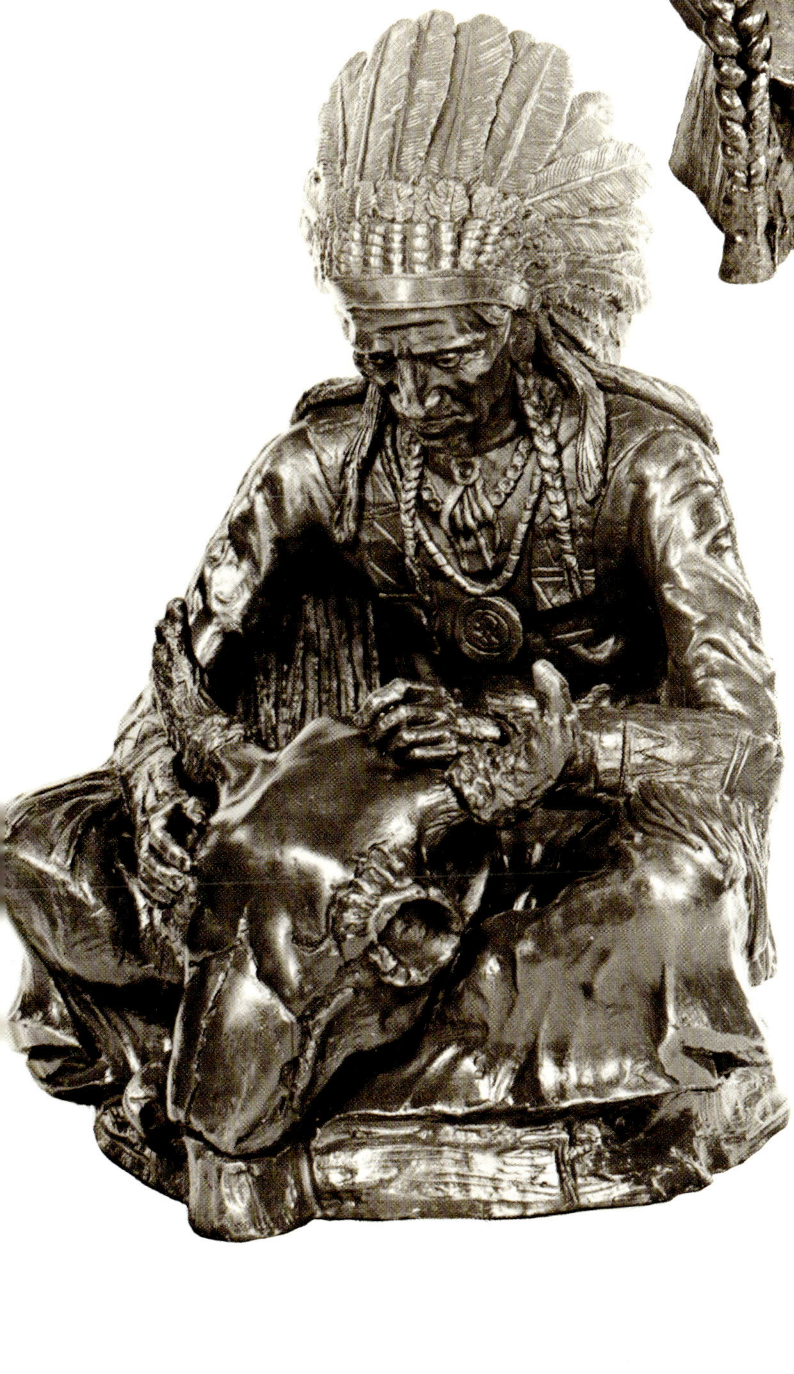

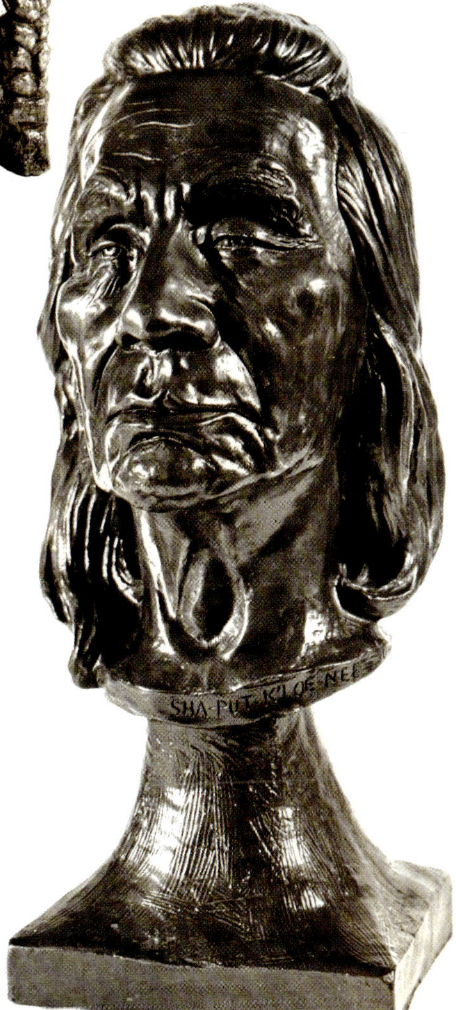

26 SHA-PUT-K'LOE-NEE,
Poker Jim. Umatilla. Bronze Height 19 inches. Dated 1934.

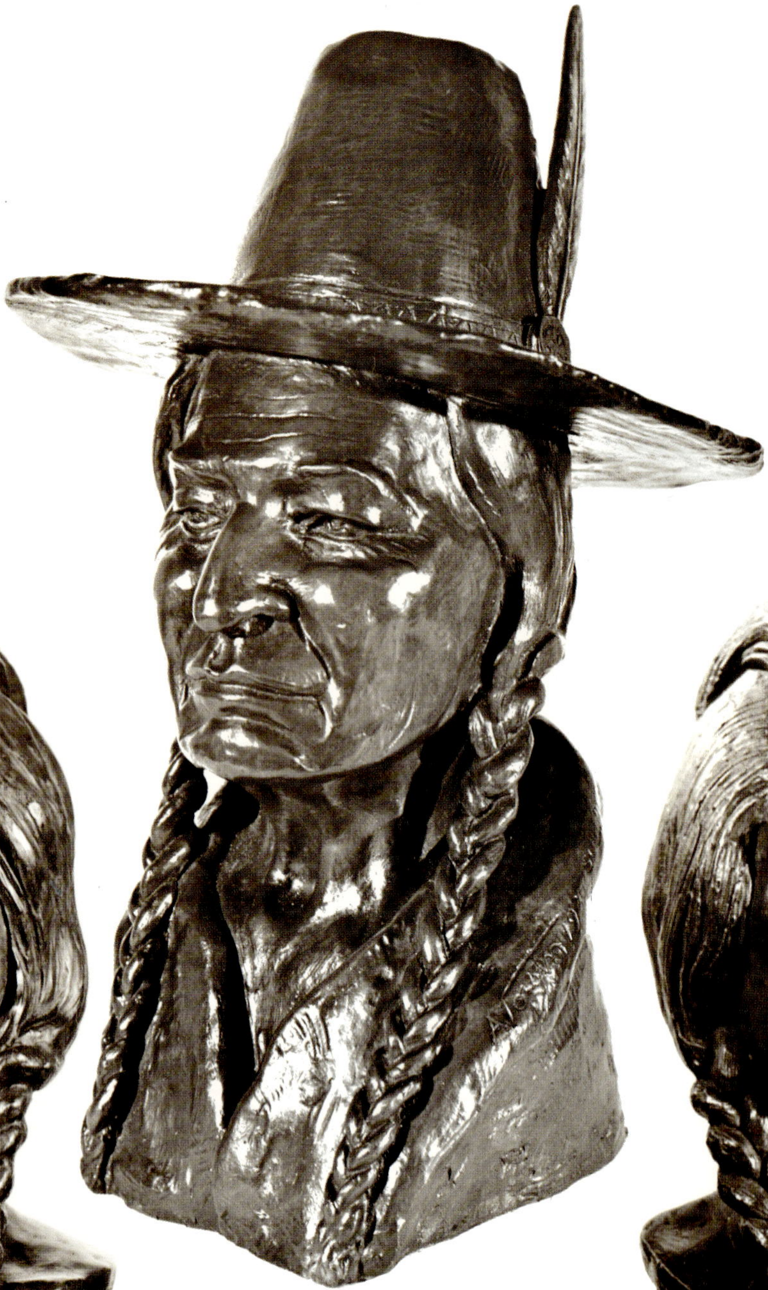

28 **PALOUSE INDIAN WITH HAT.**
Bronze Height 14½ inches. Dated 1936.

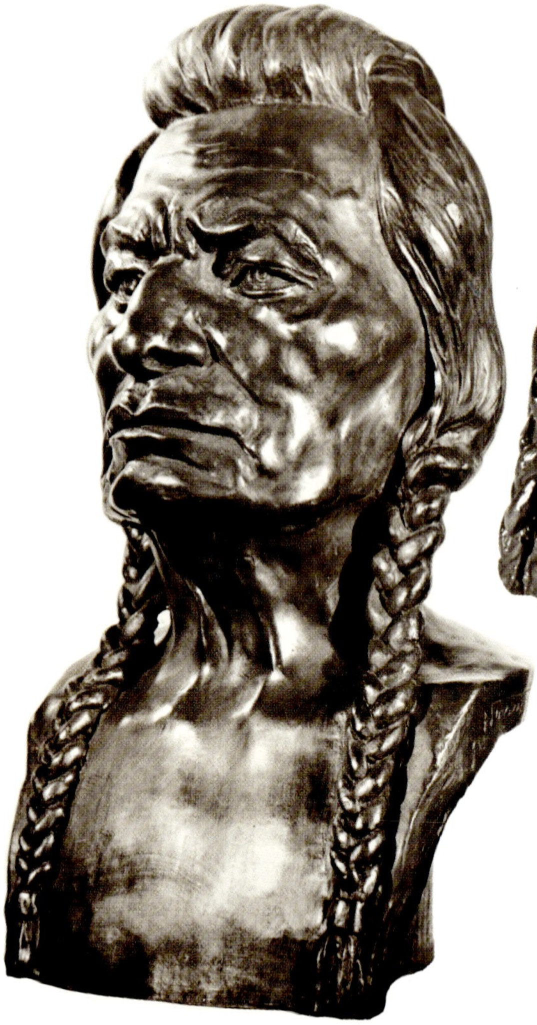

27 **SUL-LIL-SHUTE-PALOUSE,**
Father-in-Law, Poker Jim's son, Richard Burke. Bronze Height 12½ inches. Dated 1936.

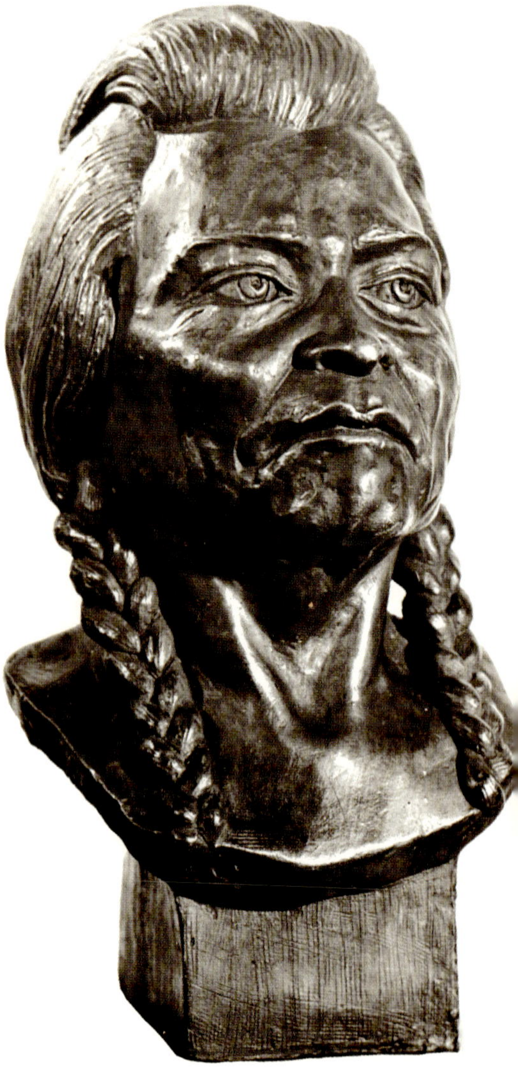

29 **CAYUSE INDIAN BUST.**
Bronze Height 11¼ inches. Dated 1935.

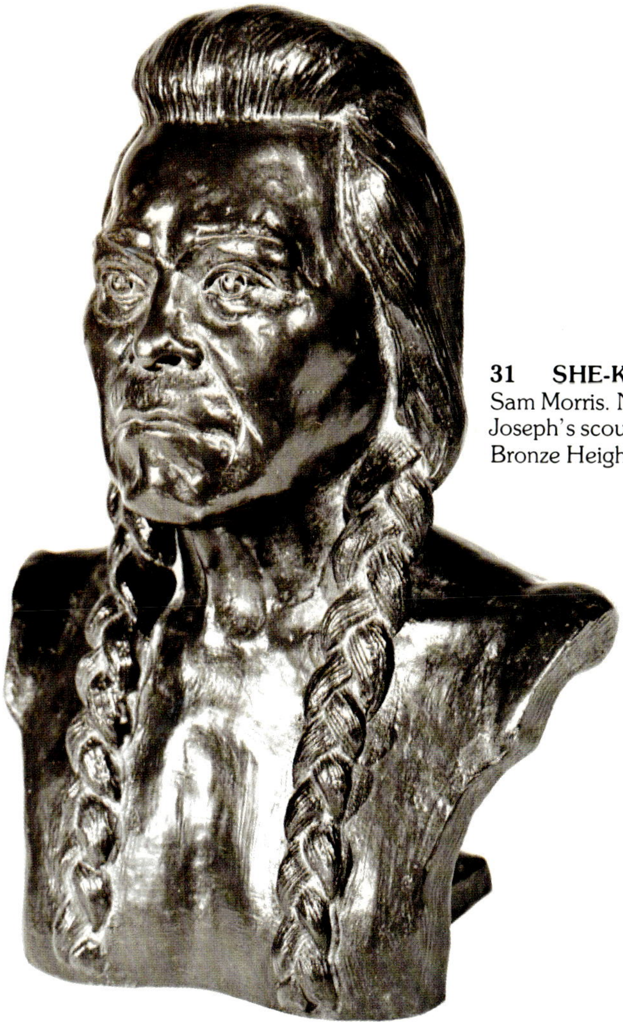

31 SHE-KAMISITS-KUN-IN,
Sam Morris. Nez Perce. One of Chief Joseph's scouts. Lapwai Indian, Idaho. Bronze Height 8½ inches. Dated 1935.

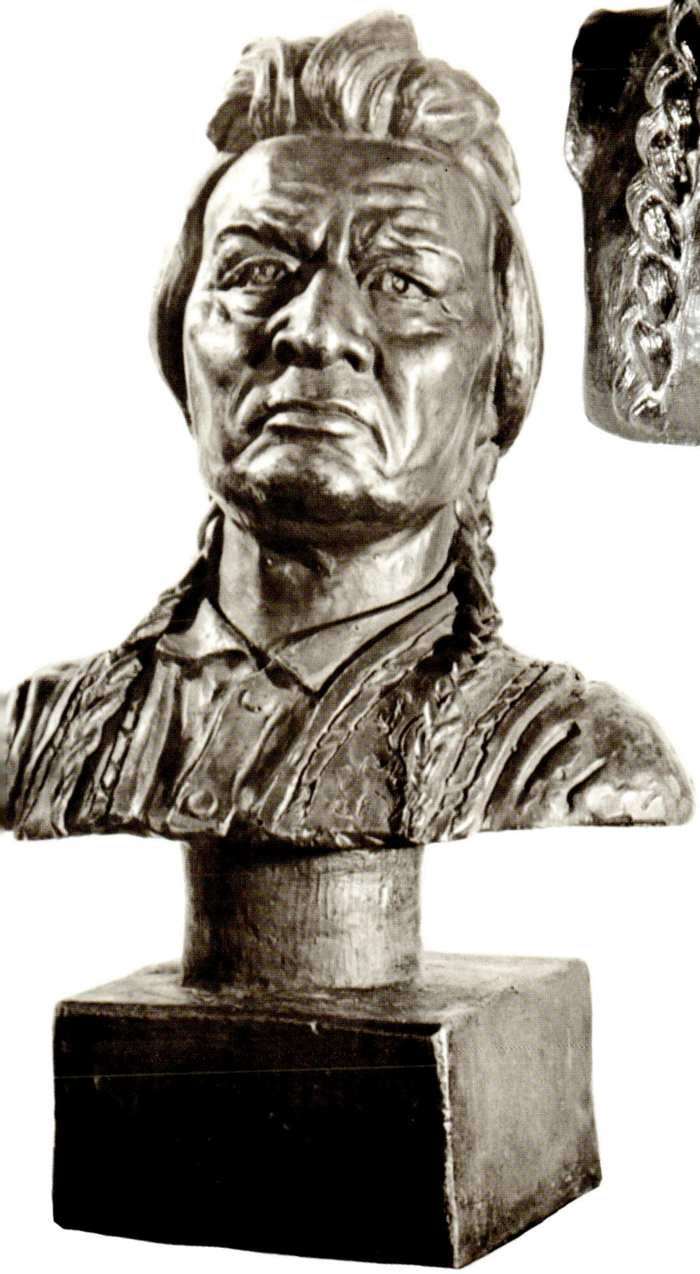

30 CHIEF BAD ROADS,
Nez Perce. Bronze Height 8 inches. Dated 1934.

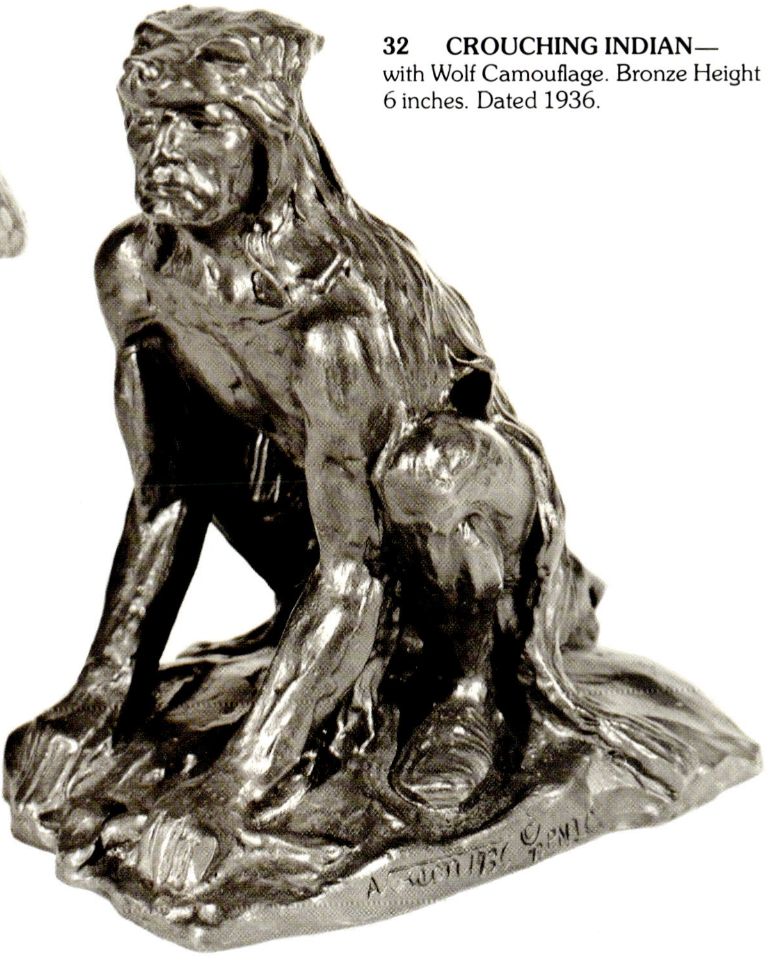

32 CROUCHING INDIAN—
with Wolf Camouflage. Bronze Height 6 inches. Dated 1936.

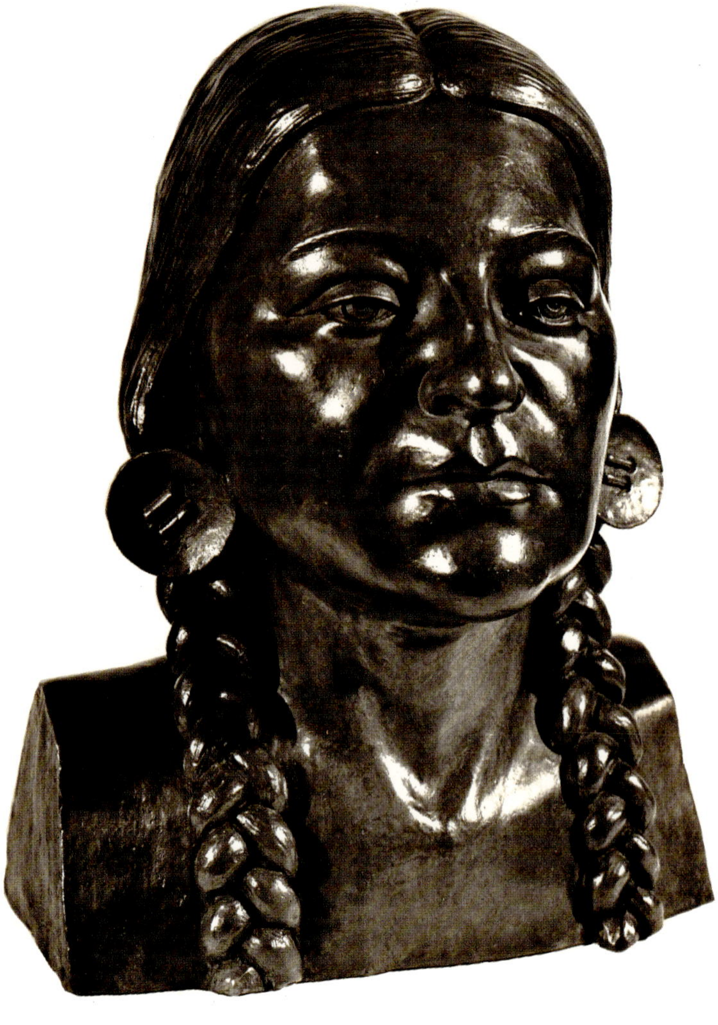

33 NEZ PERCE SQUAW.
Bronze Height 13½ inches. Dated 1936.

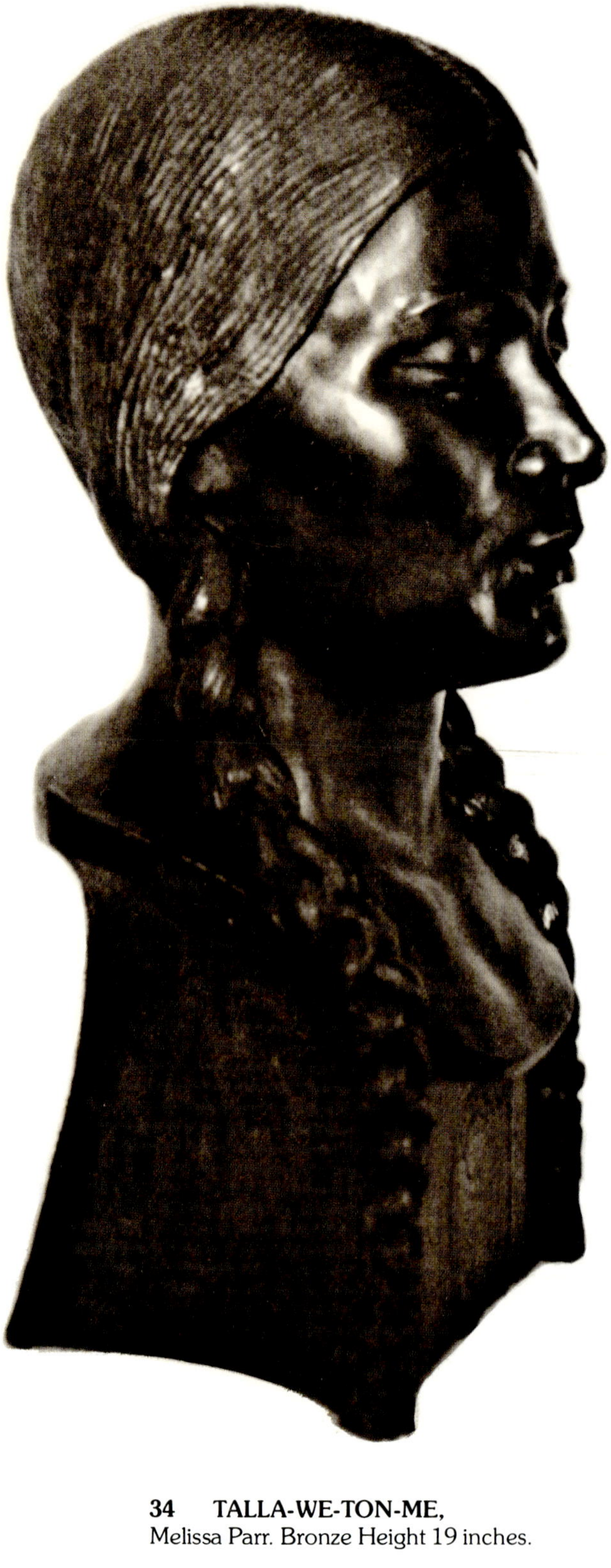

34 TALLA-WE-TON-ME,
Melissa Parr. Bronze Height 19 inches.

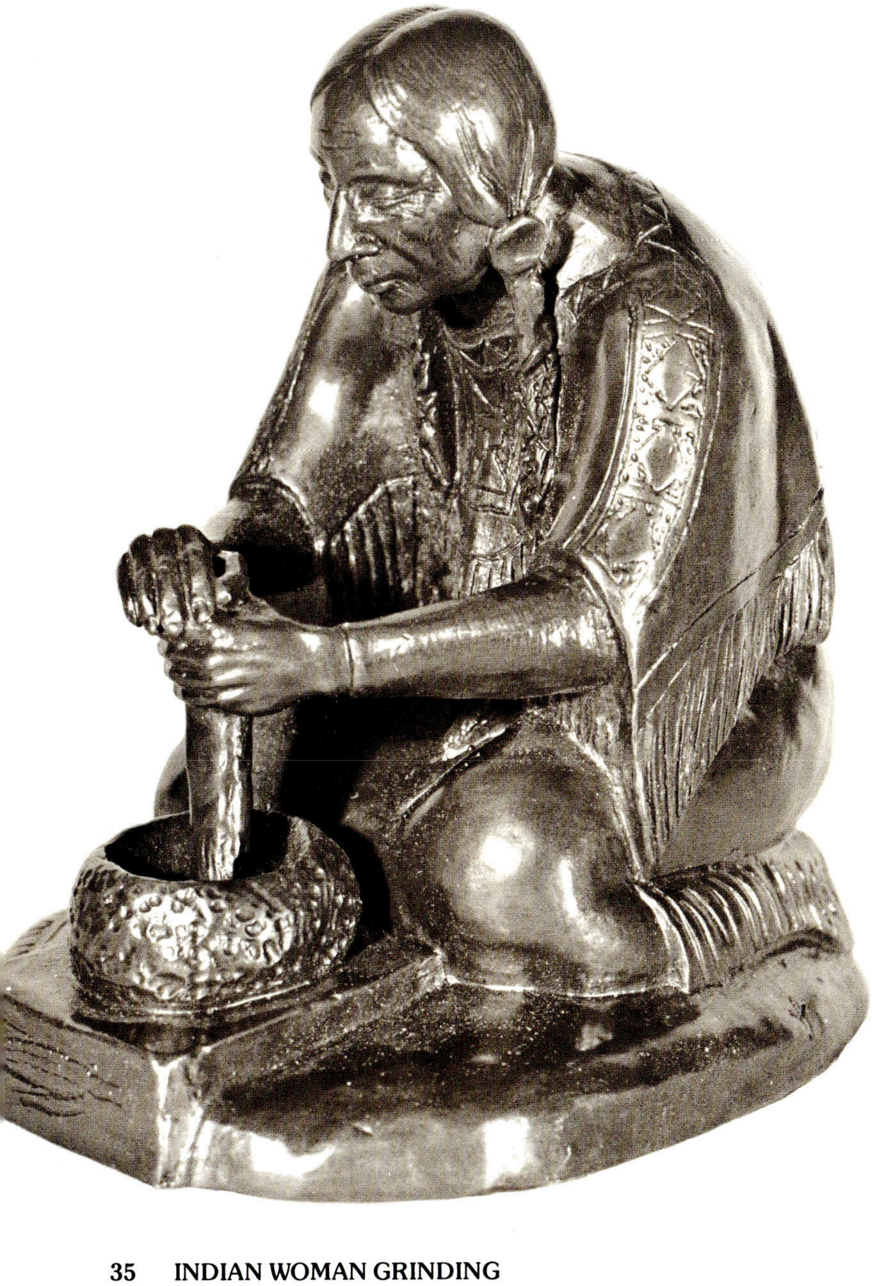

35 INDIAN WOMAN GRINDING CORN. Bronze Height 7¼ inches. Dated 1938.

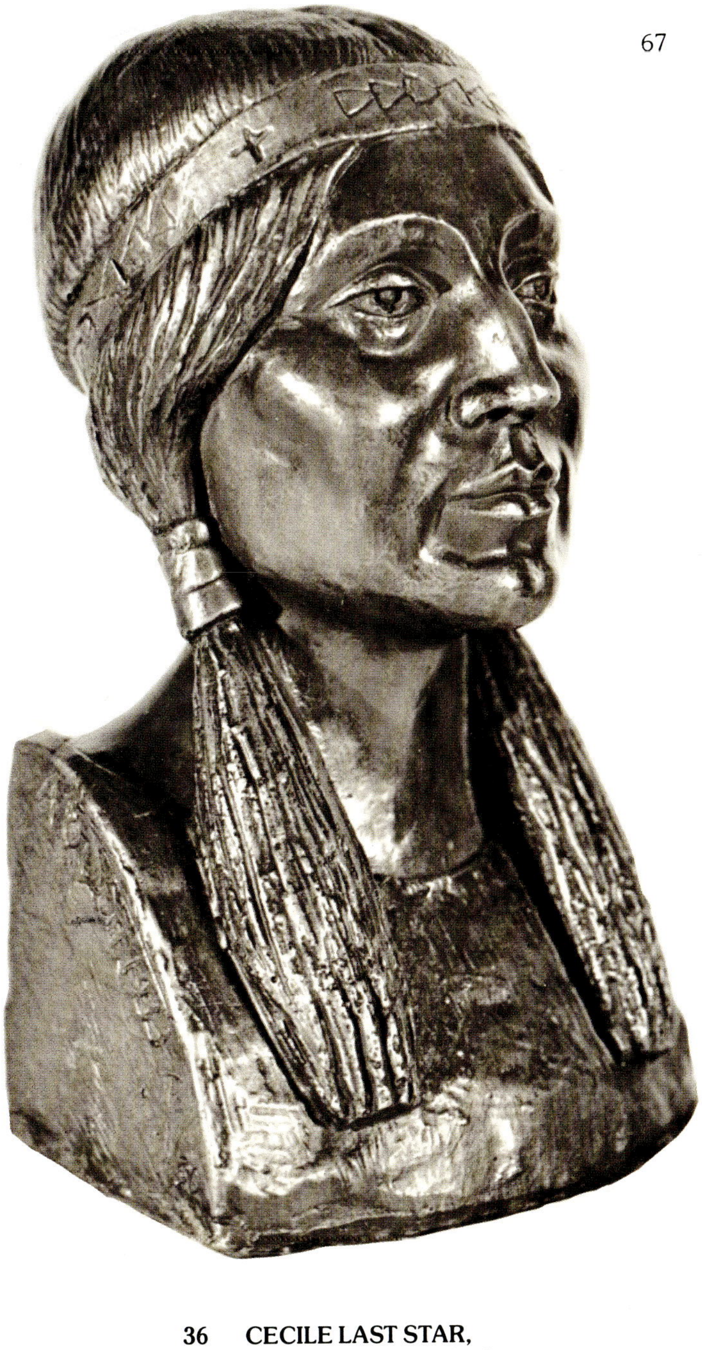

36 CECILE LAST STAR, Blackfeet Squaw. Bronze Height 6¾ inches. Dated 1933.

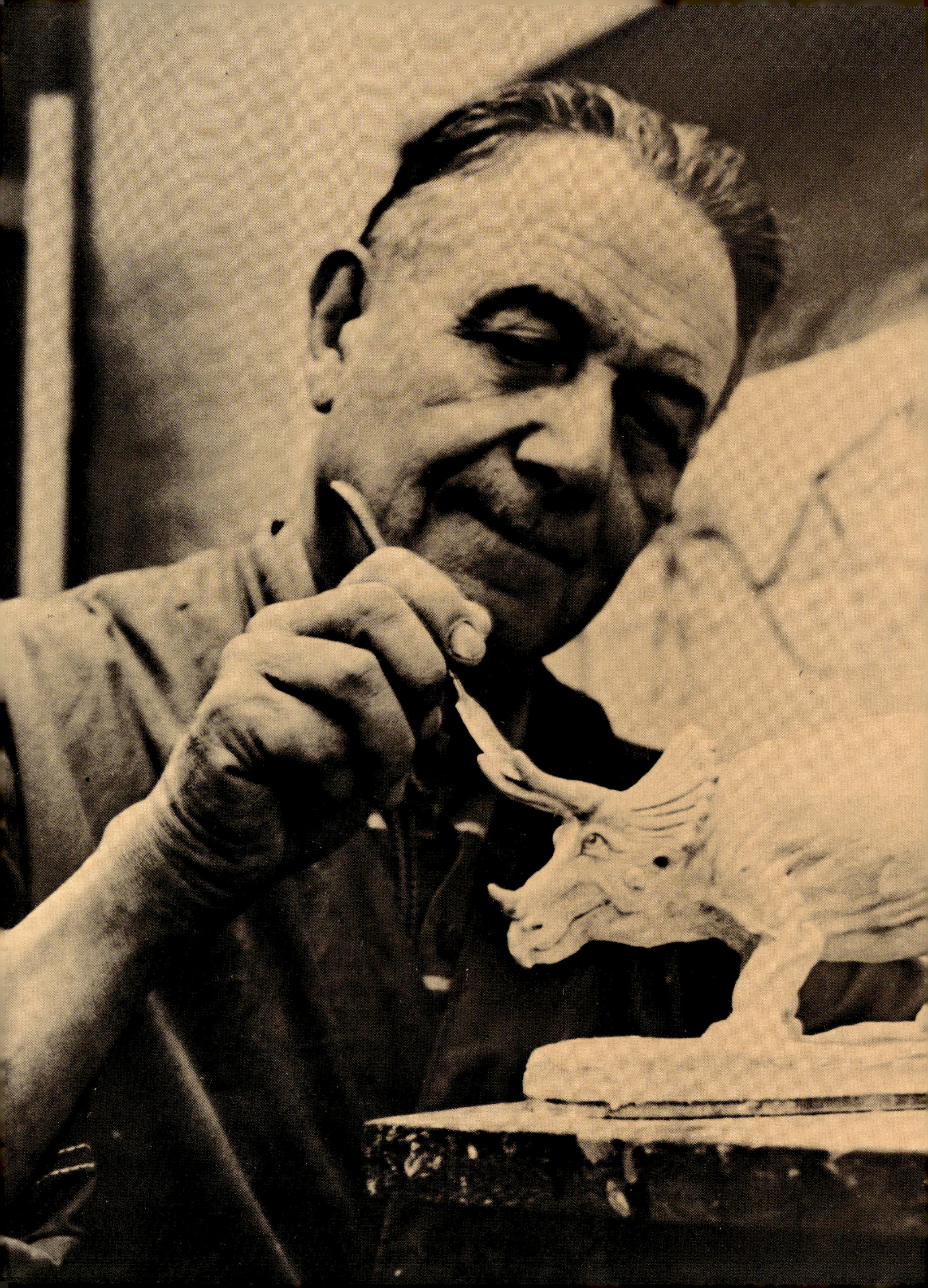

THE PREHISTORIC SERIES

DINOSAUR

In 1955, Adrien Voisin donated his time and efforts to the Steinhart Aquarium at Golden Gate Park, San Francisco. That city was interested in building a section in Anthropology, and he spent his time preparing specimens for the exhibit.

During this period, in the Old Mint Building, the United States Government was preparing bodies of dinosaurs from bones gathered for display at the Dinosaur National Monument in Uintah County, Utah. The Smithsonian Institute sent four men with drawings and plans to work with Voisin in creating six species in miniature to exact natural dimensions that could be used as models for life-size sculptures for this exhibit.

The six from this exhibit are from the original plasters made by Voisin and were never cast until commissioned by the Harmsens for this show. It was Voisin's hope that these would some day be preserved in bronze as a unit and displayed in a museum.

DINOSAUR—comes from two Greek words meaning "terrible" and "lizard," but actually they were not lizards, they were reptiles.

They existed on this planet over 100 million years B.P. (before present). No human being ever saw these strange creatures, and the last of these prehistoric reptiles were dead and gone millions of years before our earliest ancestors were known to be on this earth.

Very little was known about them. About 1820 A.D. scientists for the first time discovered fossils that had been buried for millions of years. The fossils logically explains the existence of the Dinosaur species.

Scientists, (Paleonthologists) have come up with many new theories by exhaustive studies of rock formations containing these findings, and the discovery of burial grounds which gives proof that Dinosaurs did roam the surface of the earth. Their skeleton formations have been located in many places in North America in what is now known as Montana, Wyoming, Colorado, Utah, Arizona and Texas.

These reptiles were of various sizes and shapes, and were meat eaters, plant eaters, those that lived on land, those that inhabited the water, and those that flew.

Millions of years ago the Dinosaurs died out and although scientists pursue the reasons, they conclude that climatic changes could have been the dominating factor for their disappearance. The surface of the globe has been altered many times since animal and plant life first inhabited it, and it is with this theory, that we are led to believe these strange creatures were forced out of existence because of this natural phenomenon.

Voisin putting the finishing touches on one of the original plasters of a pre-historic animal.

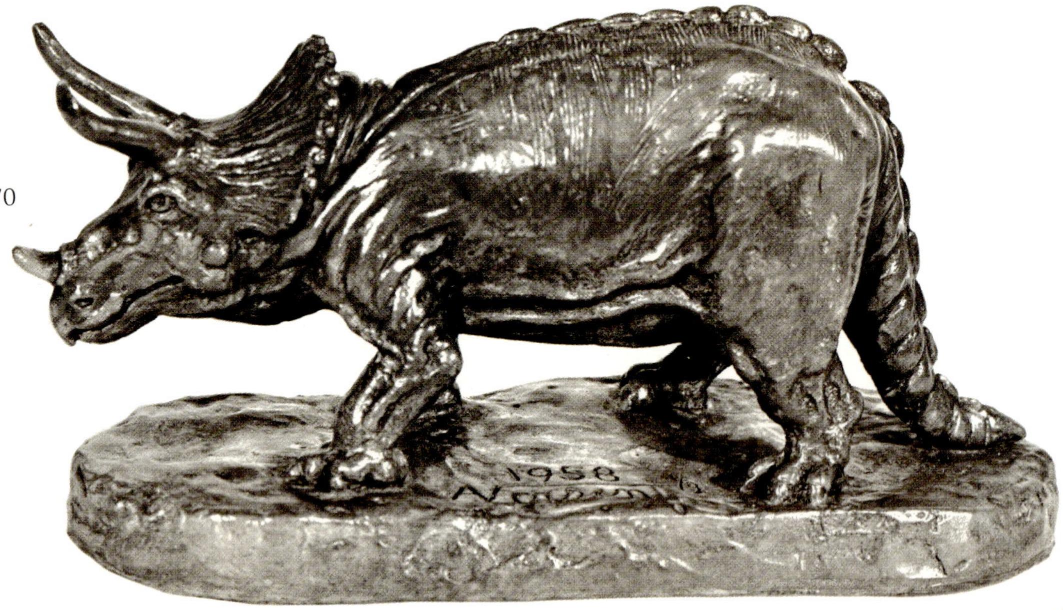

37 TRICERATOPS.
100-60 million years B.P.; fossils of triceratops, (which means 3-horned face) have been found in Wyoming and Montana. There is evidence that he could have traveled in herds. This lizard appeared in the last era of dinosaurs, the Cretaceous period. He grew to 20 feet long and probably used his horns in fighting and courting; females also grew the horns.

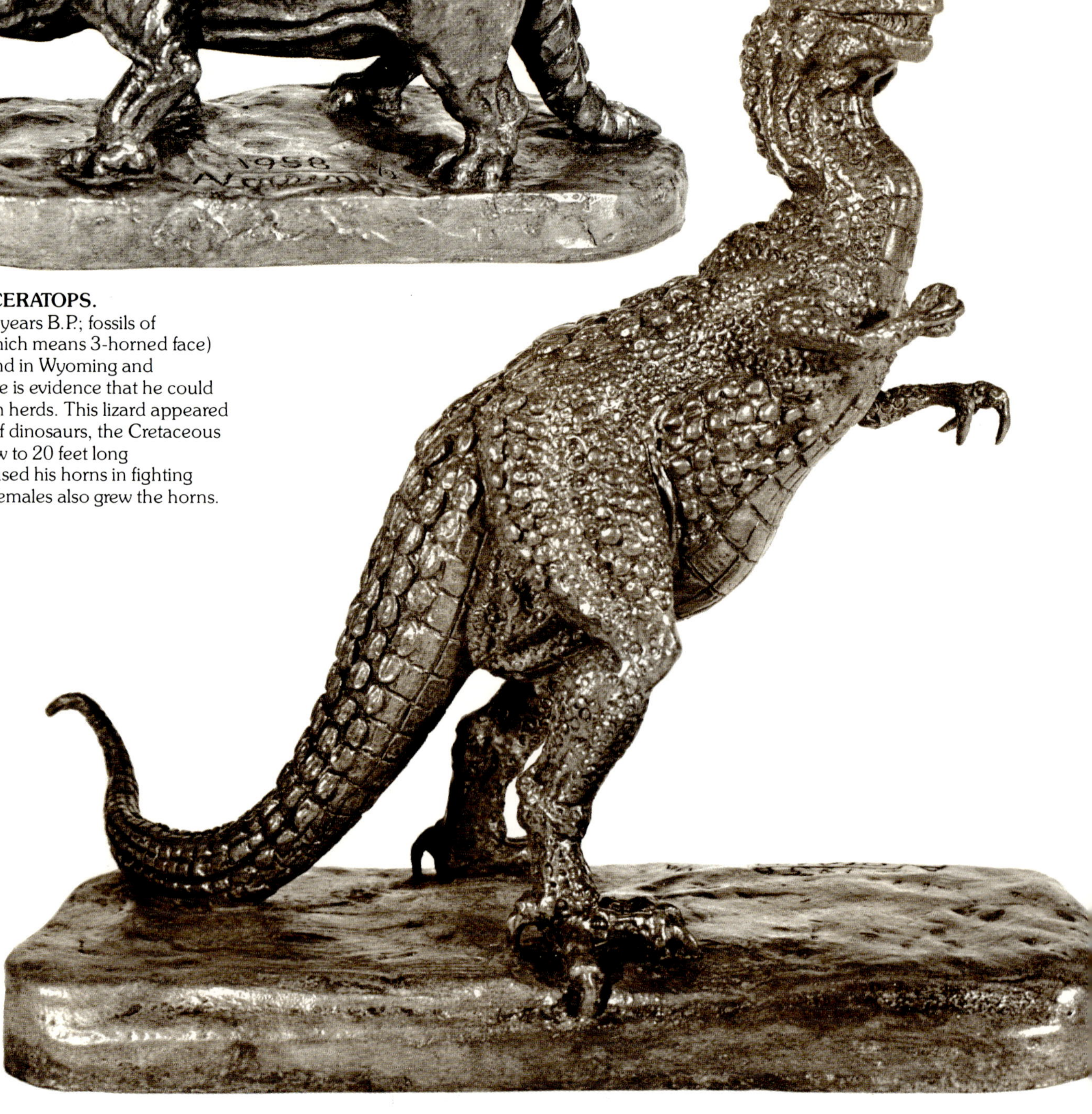

38 TYRANNOSAURUS REX.
100-60 million years B.P.; the "Tyrant Lizard King" was the most predatory reptile. Fossils have been found in South Dakota and at Hell Creek, Montana. It grew to 47 feet long and towered 20 feet above the earth. Its head measured 4′ x 3′ x 1′, and its teeth were from 3 to 6 inches long.

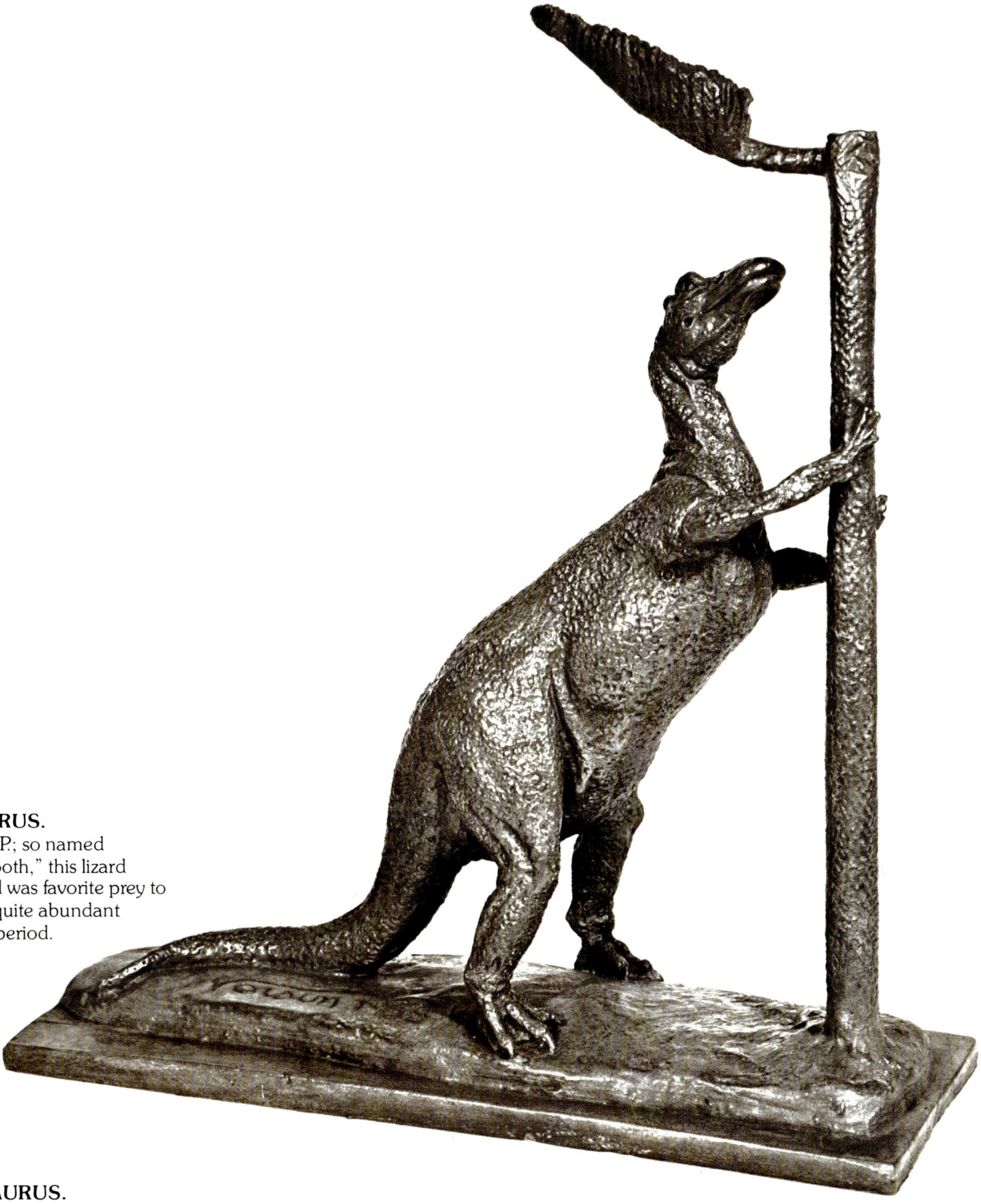

39 ANATOSAURUS.
100-60 million years B.P.; so named because of its "rough tooth," this lizard had a very soft skin and was favorite prey to most predators. It was quite abundant during the Cretaceous period.

40 BRONTOSAURUS.
100-60 million years B.P. The fossils of the "Thunder Lizard" are found in Wyoming, Montana, and other western states. This was one of the largest reptiles; it lived in the lakes and uplands that covered much of the West at that time. It was a large amphibious dinosaur, a sauropod, that was 68 feet long and weighed 20 tons.

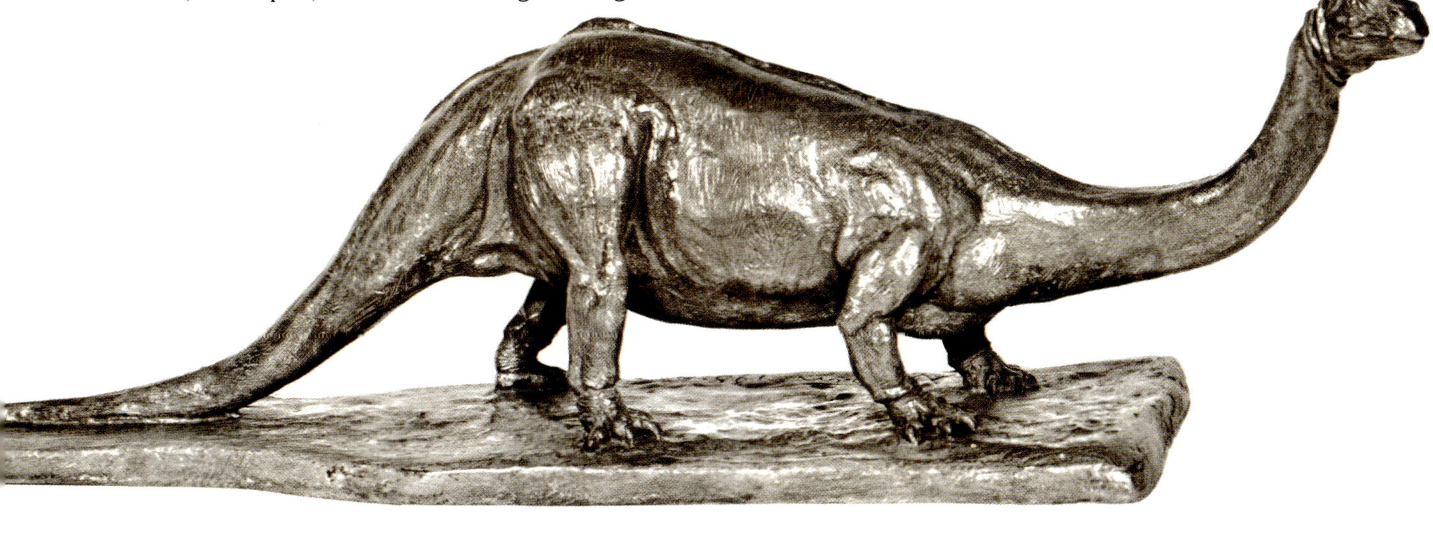

41 ANKYLOSAURUS.
80-63 million years B.P. Known as the "stiff lizard." With the exception of turtles, it was the most fully armoured reptile of all time.

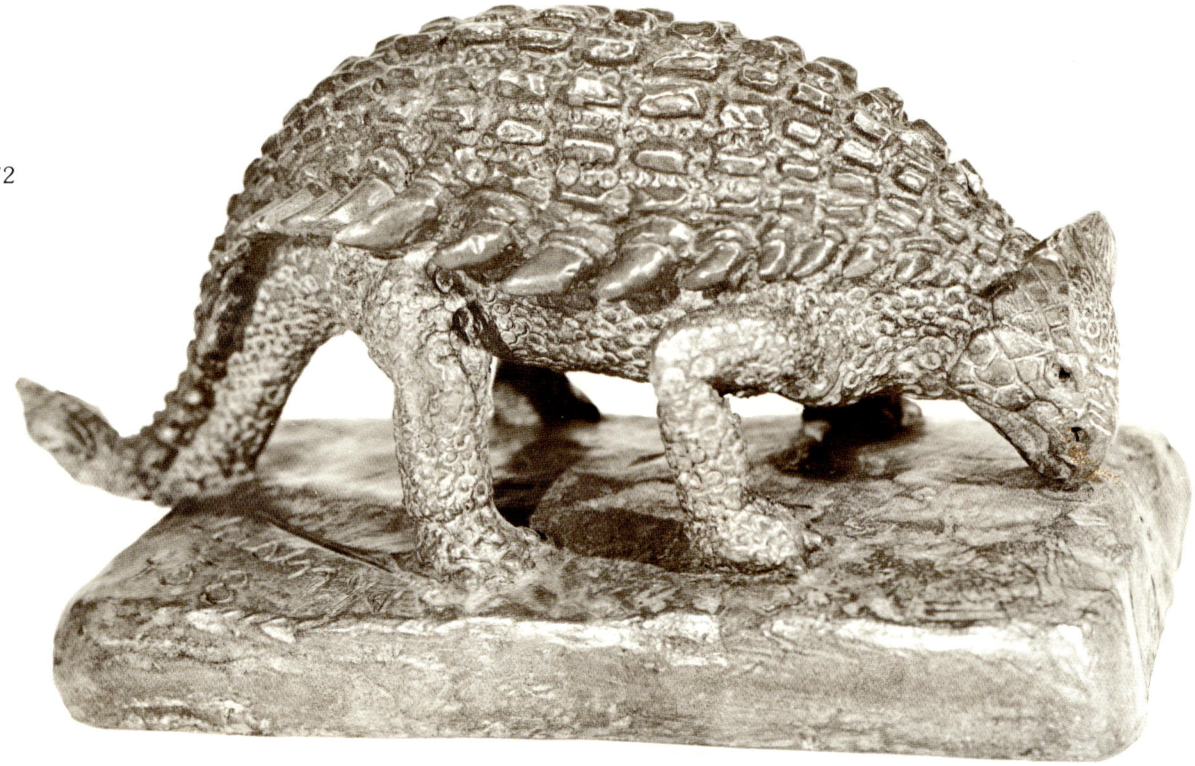

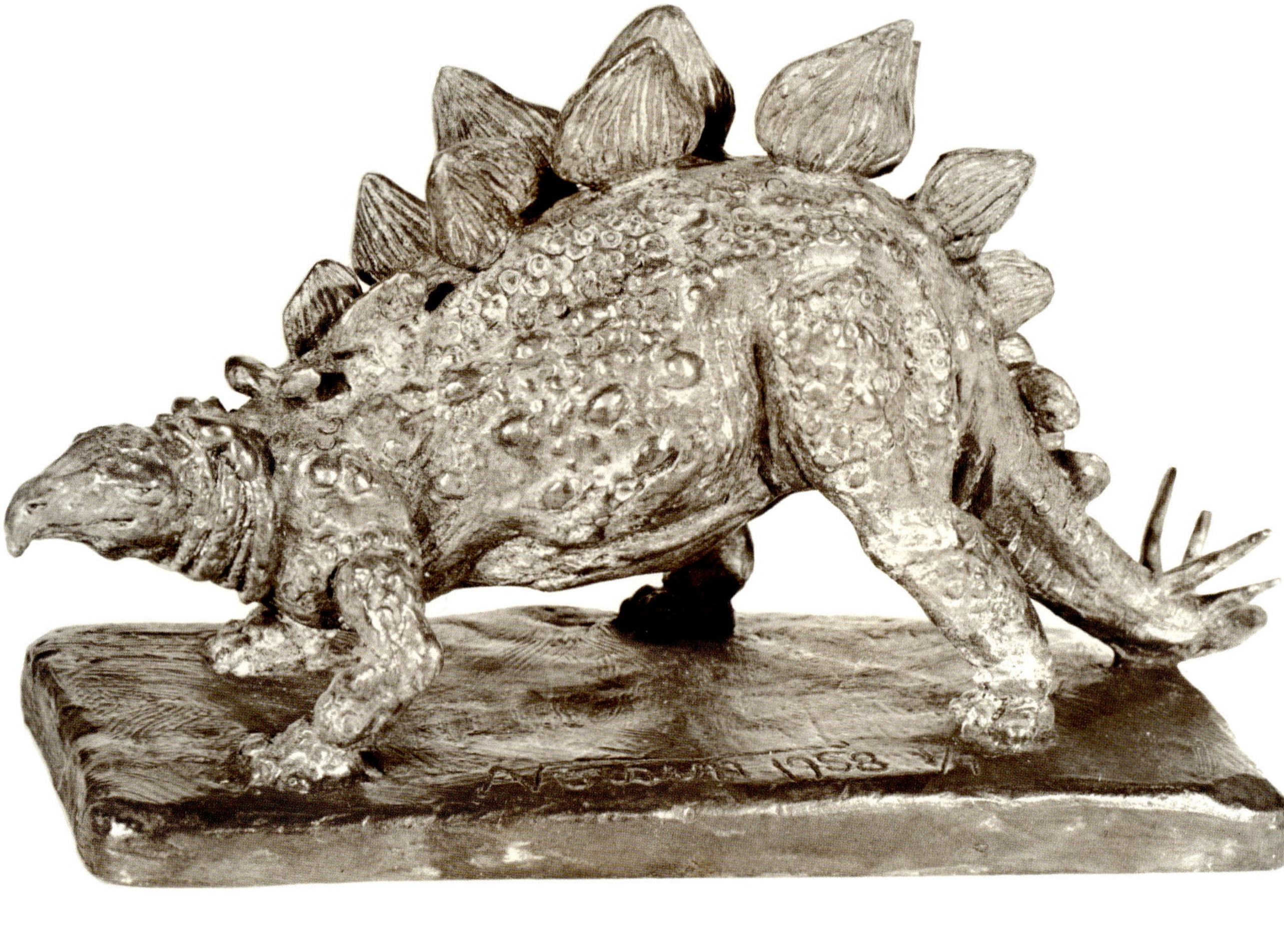

42 STEGOSAURUS.
158-110 million years B.P. The "covered lizard" was the first dinosaur to disappear, during the early Cretaceous period. It had the brain capacity of 56 CM^2 and was 14-20 feet long. It had spikes and plates that were probably used for protection and thermal regulation.

PRESERVING THE WILDERNESS

ANIMALS

Voisin was dedicated to capturing in a three-dimensional form, his friends the Indians, and realistic looking animals. His expriences as a youth, plus many visits to the San Francisco Zoo, and the Washington Park Zoo in Portland, Oregon, later aided him in sculpturing wild game in perfect proportion in lifelike action, or in reclining positions.
He apprenticed himself to a noted taxidermist for eight years, and for four years, studied woodcarving under an old French master.

All of his work displays peaceful moods. One never sees violence portrayed in his Indians or in his animals. Adrien Voisin was an ardent exponent of PEACE throughout his lifetime.

43 INDIAN CAYUSE HORSE, with Brand "DJ" on Right Side. Height 9 inches. Dated 1936.

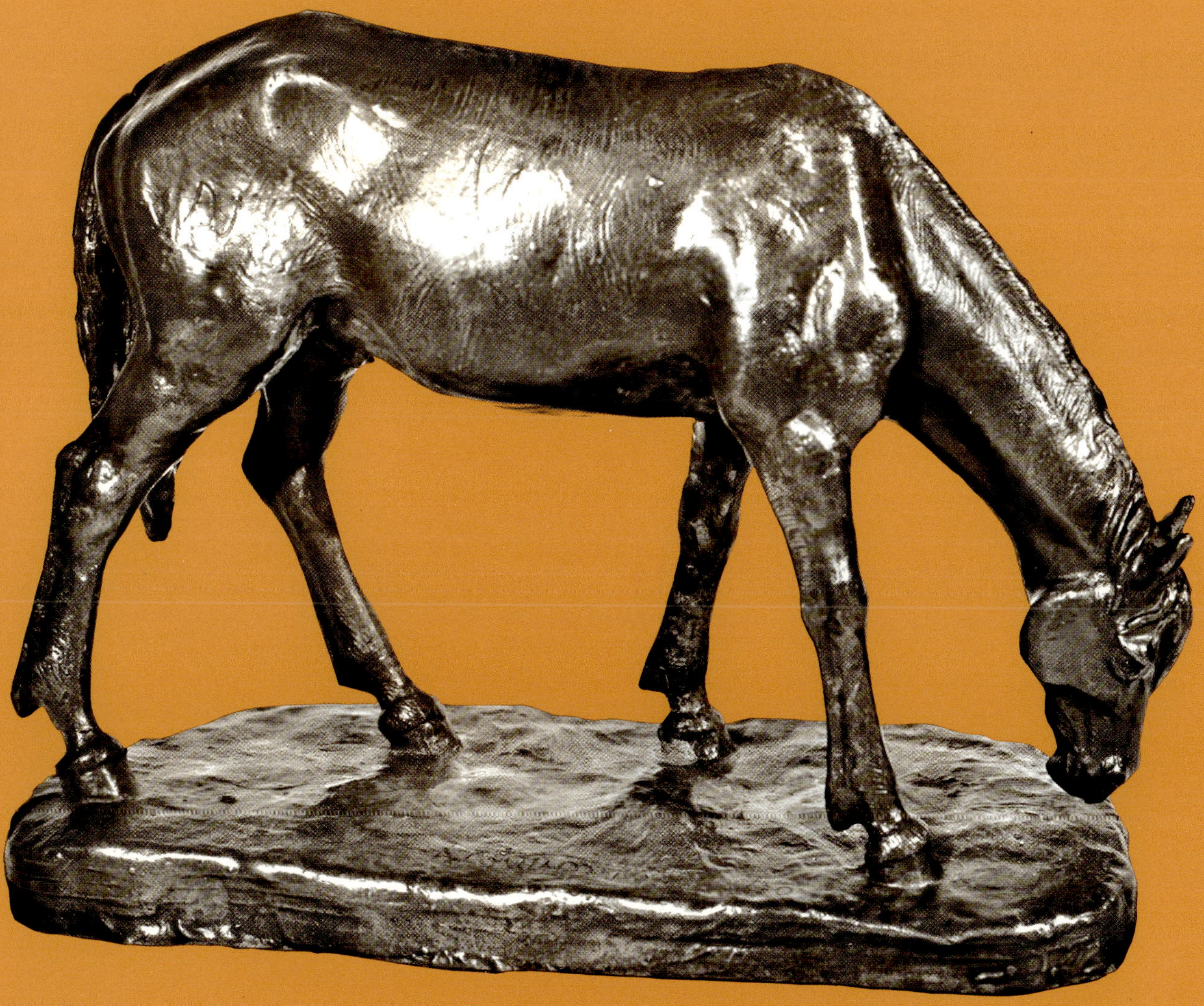

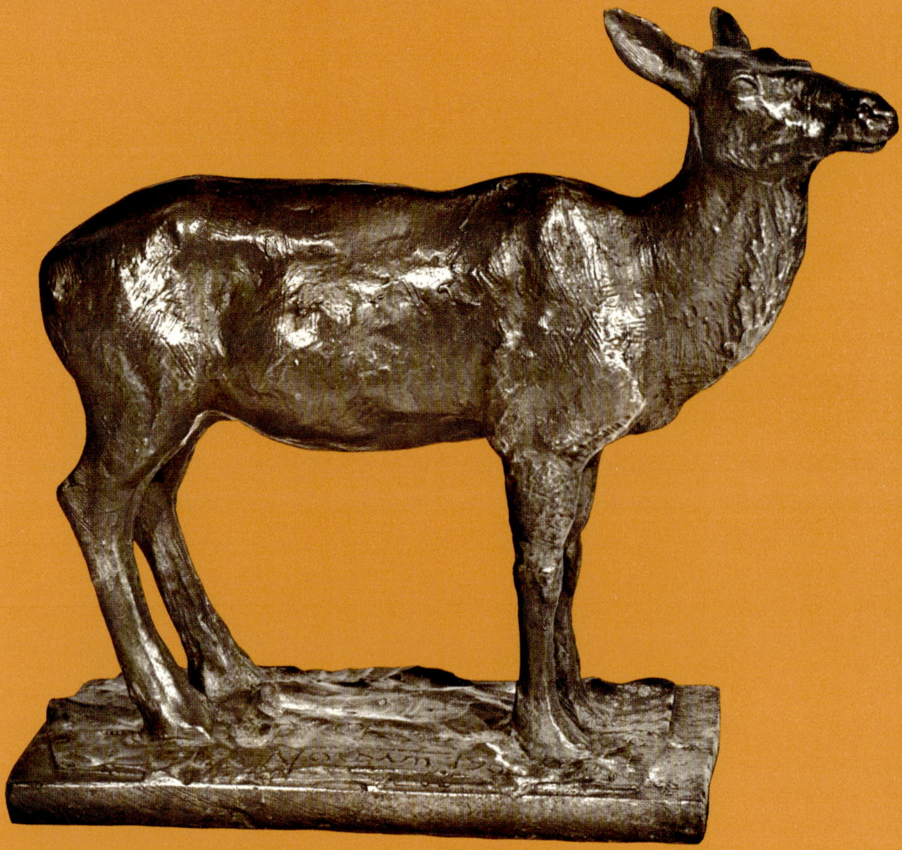

44 CALF ELK. Height 7¾ inches. Length 8¾ inches. Dated 1938.

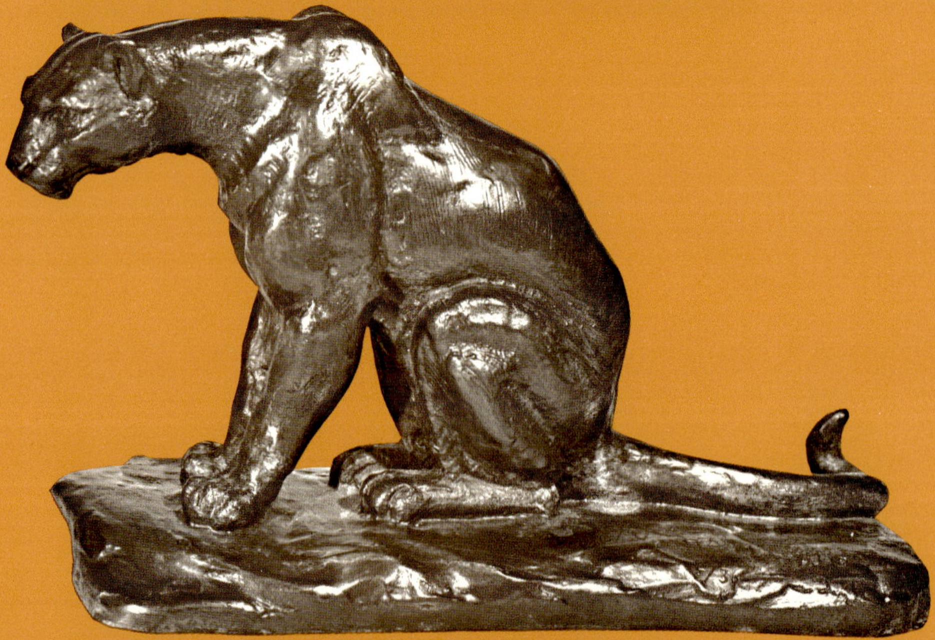

45 SITTING COUGAR. Height 6 inches. Dated 1933.

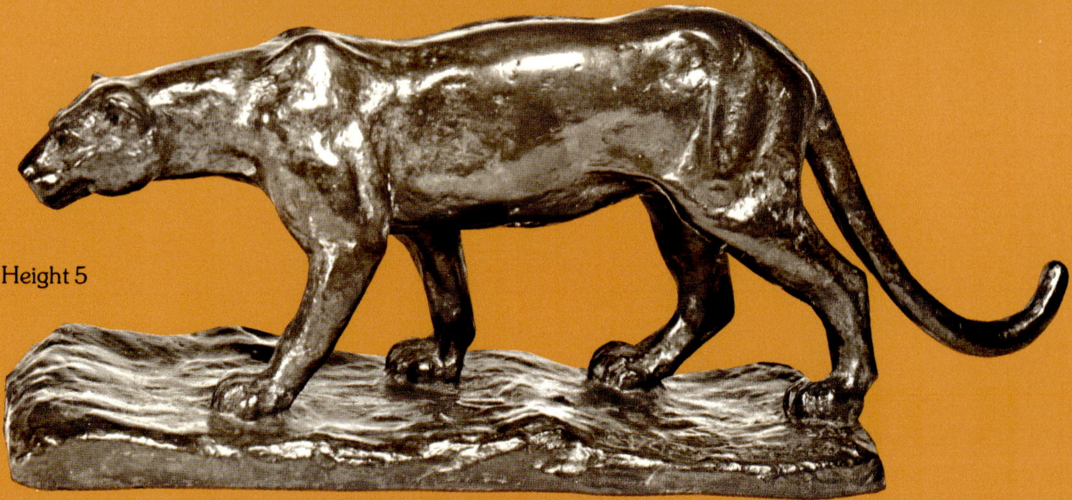

46 STALKING COUGAR. Height 5 inches. Dated 1934.

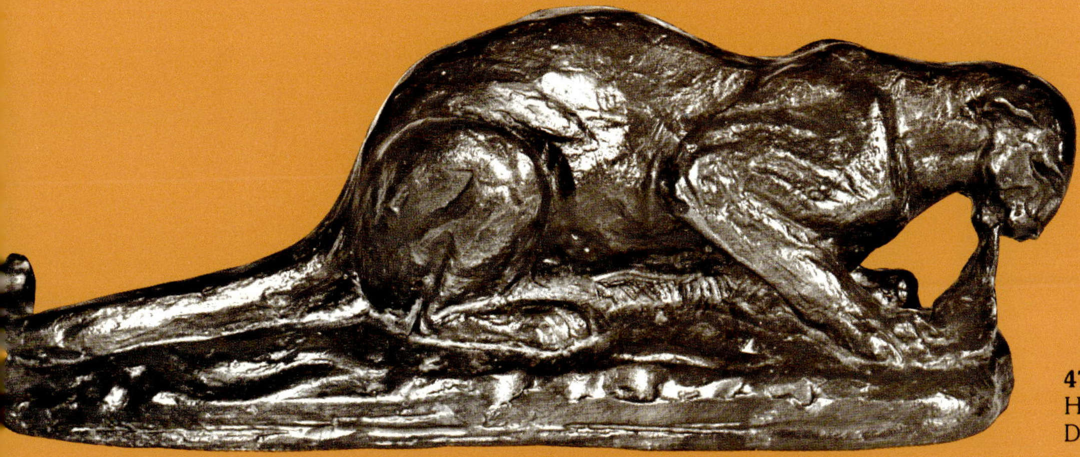

47 CROUCHING COUGAR EATING. Height 3½ inches. Length 10 inches. Dated 1938.

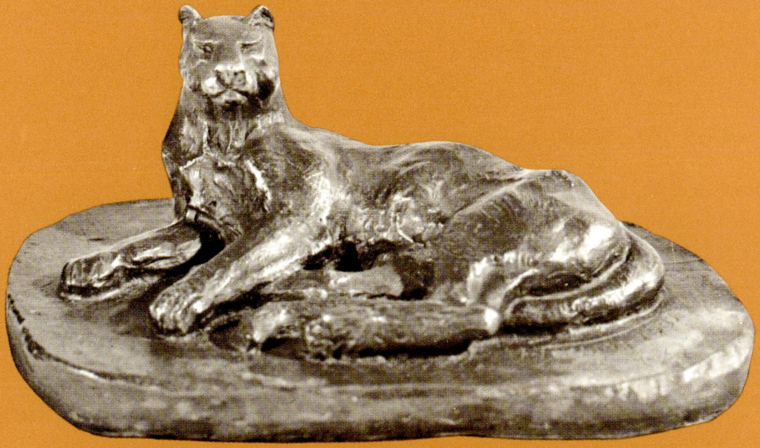

48 CONTENTED COUGAR. Length 8 inches. Dated 1935.

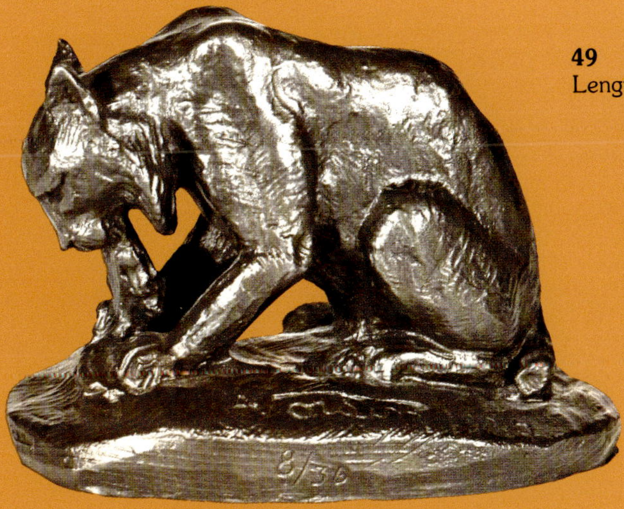

49 EATING LYNX. Height 4½ inches. Length 4½ inches. Dated 1935.

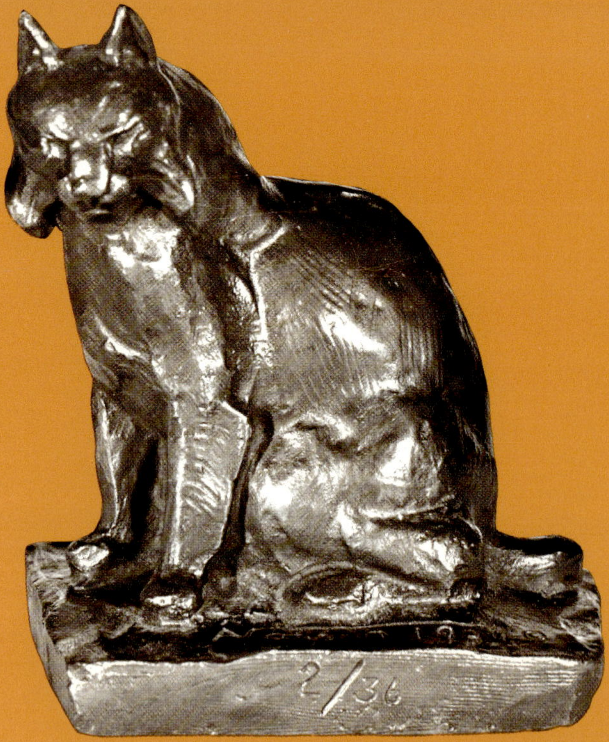

50 SITTING LYNX. Height 4½ inches. Length 6 inches. Dated 1935.

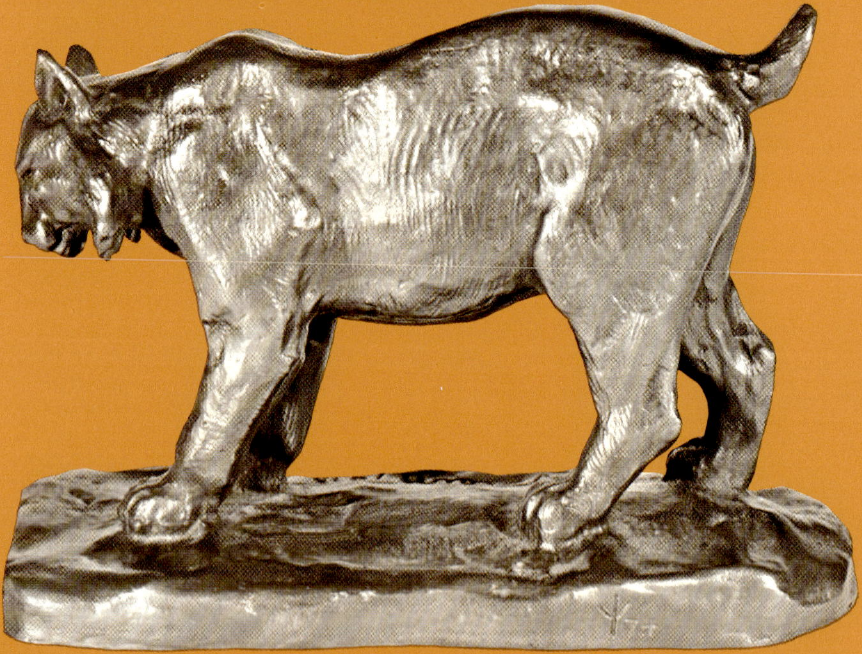

51 STANDING LYNX. Height 4¾ inches. Length 4 inches. Dated 1935.

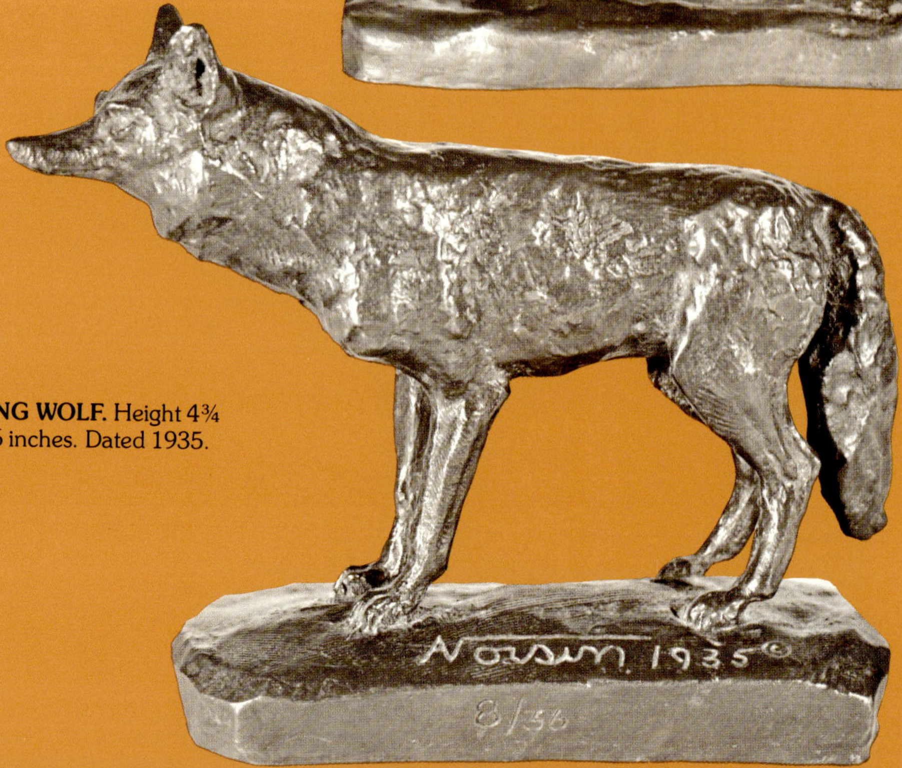

52 STANDING WOLF. Height 4¾ inches. Length 6 inches. Dated 1935.

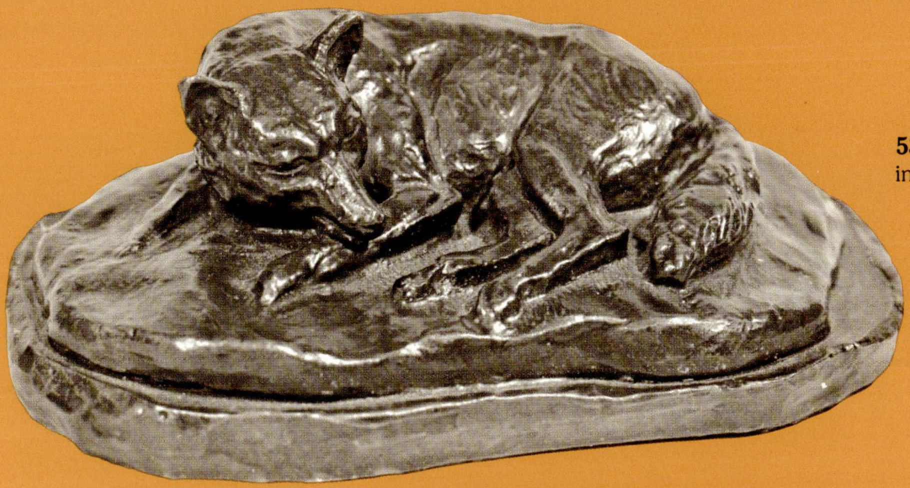

53 WOLF LYING DOWN. Height inches. Length inches. Dated 1935.

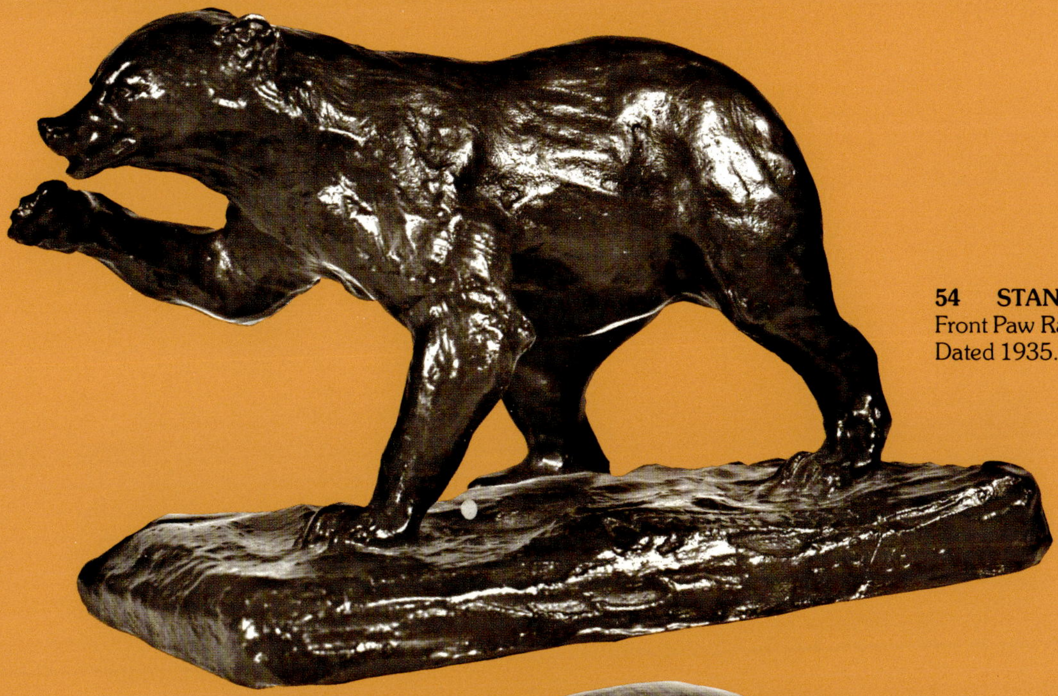

54 STANDING BEAR—with Right Front Paw Raised. Height 4½ inches. Dated 1935.

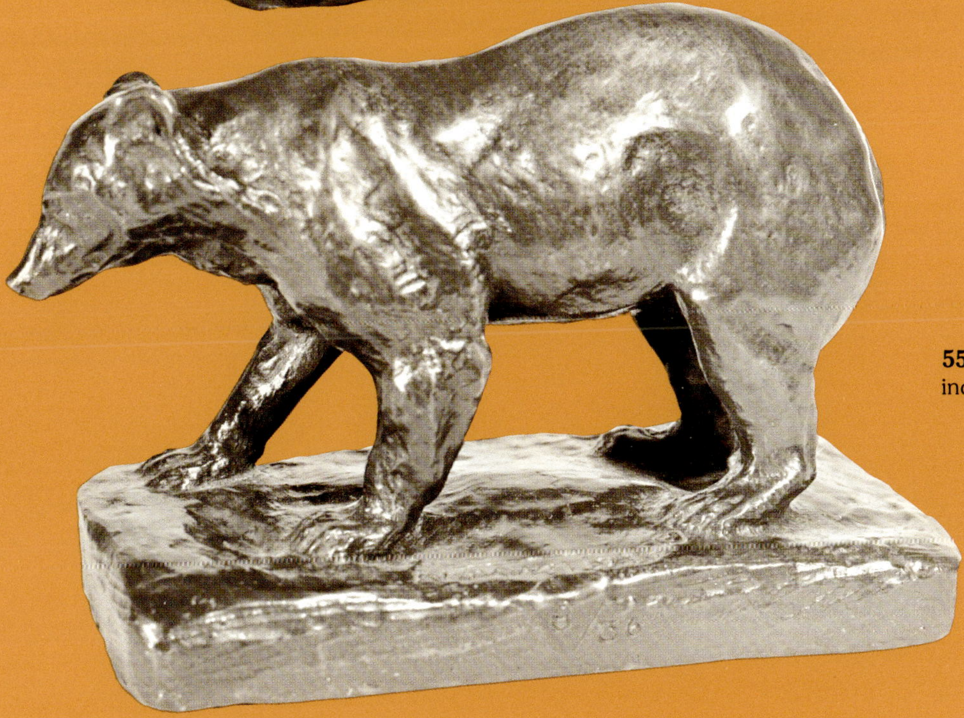

55 STANDING BEAR. Height 4 inches. Dated 1935.

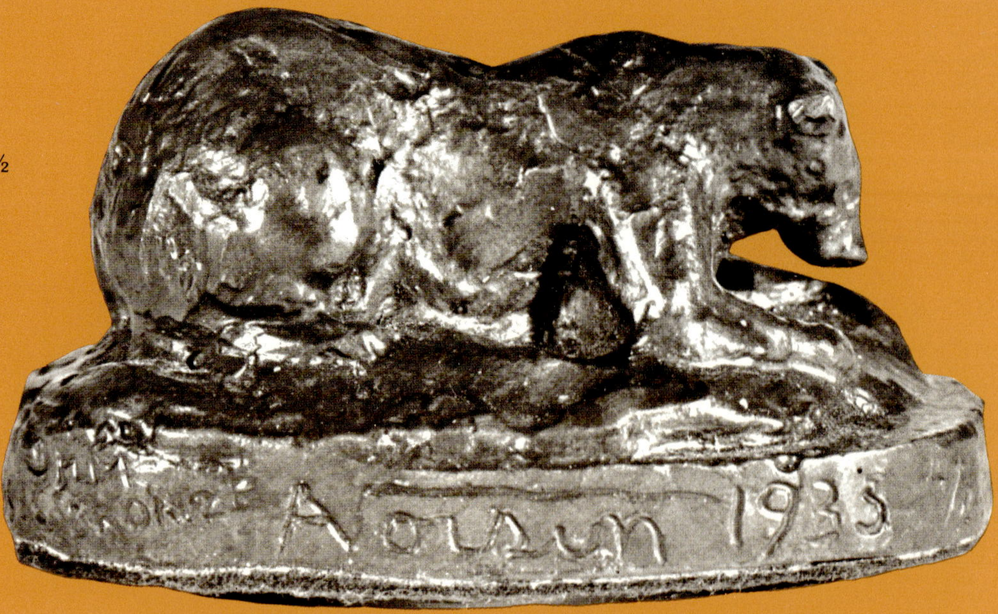

56 MINIATURE BEAR. Height 1½ inches. Dated 1933.

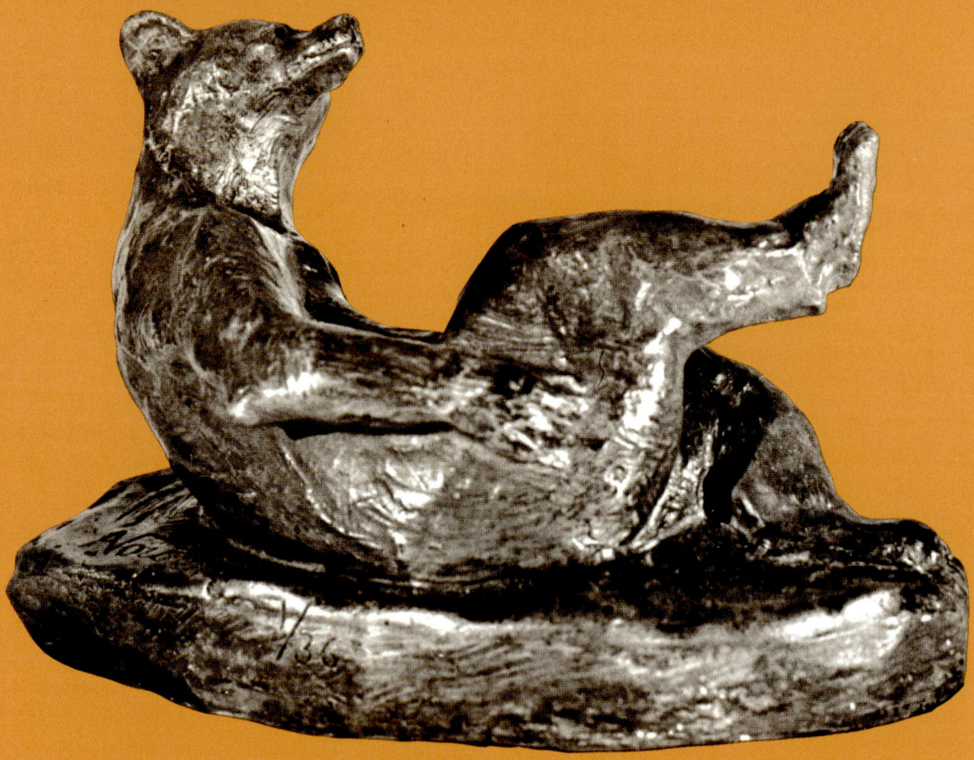

57 SCRATCHING BEAR. Height 4 inches. Length 6 inches. Dated 1935.

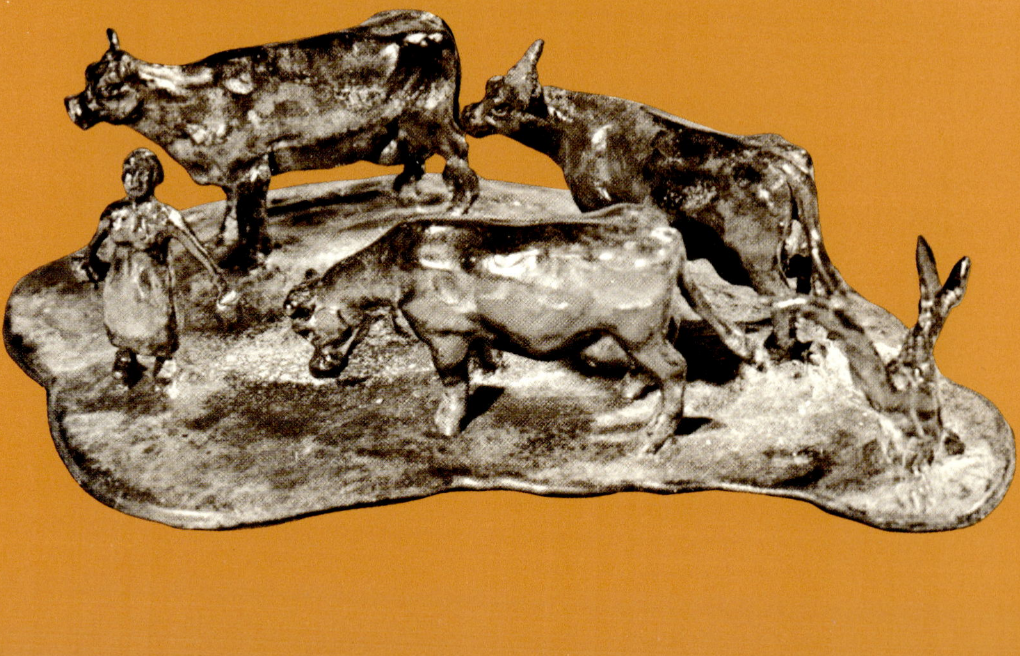

58 MILKMAID AND COWS. 1½ inches.